U.S.NRC
UNITED STATES NUCLEAR REGULATORY COMMISSION

Protecting People and the Environment

Fiscal Year 2009 Performance and Accountability Report

Brunswick Nuclear Power Plant near Southport, NC. Progress Energy oversees this facility.

MISSION

License and regulate the Nation's civilian use of

byproduct, source, and special nuclear materials

to ensure adequate protection of public health

and safety, promote the common defense and

security, and protect the environment.

Table of Contents

*This Performance and Accountability Report and previous reports are available on
the NRC Web page at http://www.nrc.gov/reading-rm/doc-collections/nuregs/staff/sr1542/*

U.S. Nuclear Regulatory Commission Commissioner Dale E. Klein, Chairman Gregory B. Jaczko, and Commissioner Kristine L. Svinicki

The FY 2009 Performance and Accountability Report provides performance results and audited financial statements that enable Congress, the President, and the public to assess the performance of the agency in achieving its mission and stewardship of its resources. The report contains a concise overview, Management's Discussion and Analysis, as well as performance and financial sections. Details of performance results and program evaluations can be found in the Other Accompanying Information section.

A Message from the Chairman

I am pleased to present the U.S. Nuclear Regulatory Commission (NRC) Performance and Accountability Report for fiscal year (FY) 2009. The report provides key financial and performance information to Congress and the American people. Continuing our trend of excellence in reporting, the NRC received an eighth Certificate of Excellence in Accountability Reporting from the Association of Government Accountants (AGA) for our FY 2008 Performance and Accountability Report.

Our mission of protecting public health and safety, promoting common defense and security, and protecting the environment is critical both to the licensees we regulate and to the public we serve. This reports highlights our achievements in meeting our mission through the agency's two strategic goals of safety and security, while adhering to the principles of good regulation—independence, openness, efficiency, clarity, and reliability.

In FY 2009, while the NRC maintained effective and efficient oversight of 104 nuclear power plants through emphasis on strengthening the interrelationship among safety, security, and emergency preparedness, the agency concurrently continued to review the critical safety aspects of new reactor designs, environmental siting, and licensing of new nuclear power plants. The NRC also continued to focus on the safe and secure use of nuclear materials through effective oversight of fuel facilities, uranium recovery sites, decommissioning sites, and nuclear material user licensees. In addition, the agency reviewed new applications, including those for uranium enrichment facilities and uranium recovery, to assure that public health and safety and the environment would be protected.

Commensurate with the NRC's programmatic achievements is a commitment to prudently manage the resources entrusted to it by the American public. The NRC continues to evaluate its internal controls and to implement internal control improvements, including those related to financial reporting and financial management systems, as required by the Federal Managers Financial Integrity Act (FMFIA) and Federal Information Security Management Act (FISMA). Based on the FMFIA assessments, I have concluded that there is reasonable assurance that the NRC is in substantial compliance with the FMFIA and FISMA. The NRC is pleased to have obtained an unqualified opinion on the agency's financial statements for the sixth consecutive year. This report demonstrates that the agency's financial and performance data are reliable and complete.

The NRC is proud of this year's performance in achieving the agency's safety and security goals and looks forward to continuing its high-quality service to the American public in FY 2010 and beyond.

Gregory B. Jaczko
Chairman
November 13, 2009

AGA.

CERTIFICATE OF EXCELLENCE IN ACCOUNTABILITY REPORTING®

Presented to the

U.S. Nuclear Regulatory Commission

In recognition of your outstanding efforts preparing NRC's Performance and Accountability Report for the fiscal year ended September 30, 2008.

A *Certificate of Excellence in Accountability Reporting* is presented by AGA to federal government agencies whose annual Performance and Accountability Reports achieve the highest standards demonstrating accountability and communicating results.

John H. Hummel, CGFM
Chair, Certificate of Excellence
in Accountability Reporting Board

Relmond P. Van Daniker, DBA, CPA
Executive Director, AGA

Chapter 1

Management's Discussion and Analysis

Watts Bar nuclear power plant is located just south of Watts Bar Reservoir on the Tennessee River near Spring City in east Tennessee. It is Tennessee Valley Authority's third nuclear power plant.

The U.S. Nuclear Regulatory Commission (NRC) headquarters

Introduction

The U.S. Nuclear Regulatory Commission (NRC) Performance and Accountability Report presents the agency's program performance and financial management information during fiscal year (FY) 2009. The annual report provides an opportunity for the public to assess how effectively the NRC uses its funds to achieve results. When preparing this report, the NRC staff followed the requirements of the Chief Financial Officers Act, as amended by the Reports Consolidation Act, Government Management Reform Act of 1994, and Government Performance Results Act of 1993. This Performance and Accountability Report covers activities from October 1, 2008, to September 30, 2009.

The NRC emphasizes keeping the public informed of its activities. Visit our Web site at http://www.nrc.gov to access this report and to learn more about who we are and what we do to serve the American public.

Chapter 1, "Management's Discussion and Analysis," provides an overview of the NRC and its accomplishments during FY 2009. Chapter 1 consists of the following six sections: "About the NRC" describes the agency's mission, organizational structure, and regulatory responsibility; "Program Performance Overview" summarizes the agency's success in achieving its strategic goals, which are further described in Chapter 2; "Program Performance Results" outlines the results of the agency's program performance; "Future Challenges" includes forward-looking information; "Financial Performance Overview" highlights the NRC's financial position and audit results contained in Chapter 3; and "Systems, Controls, and Legal Compliance" describes the agency's compliance with key legal and regulatory requirements.

About the NRC

The U.S. Congress established the NRC on January 19, 1975, as an independent Federal agency regulating the commercial and institutional uses of nuclear materials. The Atomic Energy Act, as amended, and the Energy Reorganization Act, as amended, define the NRC's purpose. These acts provide the foundation for the NRC's mission to regulate the Nation's civilian use of byproduct, source, and special nuclear materials to ensure adequate protection of public health and safety, to promote the common defense and security, and to protect the environment.

The agency regulates civilian nuclear power plants and other nuclear facilities, as well as other uses of nuclear materials. These other uses include nuclear medicine programs at hospitals; academic activities at educational institutions; research work; industrial applications, such as gauges and testing equipment; and the transport, storage, and disposal of nuclear materials and wastes.

To fulfill its responsibility to protect public health and safety, the NRC performs the following three principal regulatory functions:

1. establishes standards and regulations

2. issues licenses for nuclear facilities and users of nuclear materials

3. inspects facilities and users of nuclear materials to ensure compliance with regulatory requirements

Organization

The NRC is headed by a Commission composed of five members, with one member designated by the President to serve as Chairman. With the advice and consent of the U.S. Senate, the President appoints each member to serve a 5-year term. The Chairman is the principal executive officer and official spokesman for the Commission. The Executive Director for Operations carries out the Commission's program policies and decisions.

The NRC's headquarters is located in Rockville, MD. Four regional offices are located in King of Prussia, PA; Atlanta, GA; Lisle, IL; and Arlington, TX. The NRC's technical training center is located in Chattanooga, TN. The NRC also employs at least two resident inspectors at each of the Nation's nuclear power reactor sites. The NRC's Operations Center, located at the headquarters building in Rockville, MD,

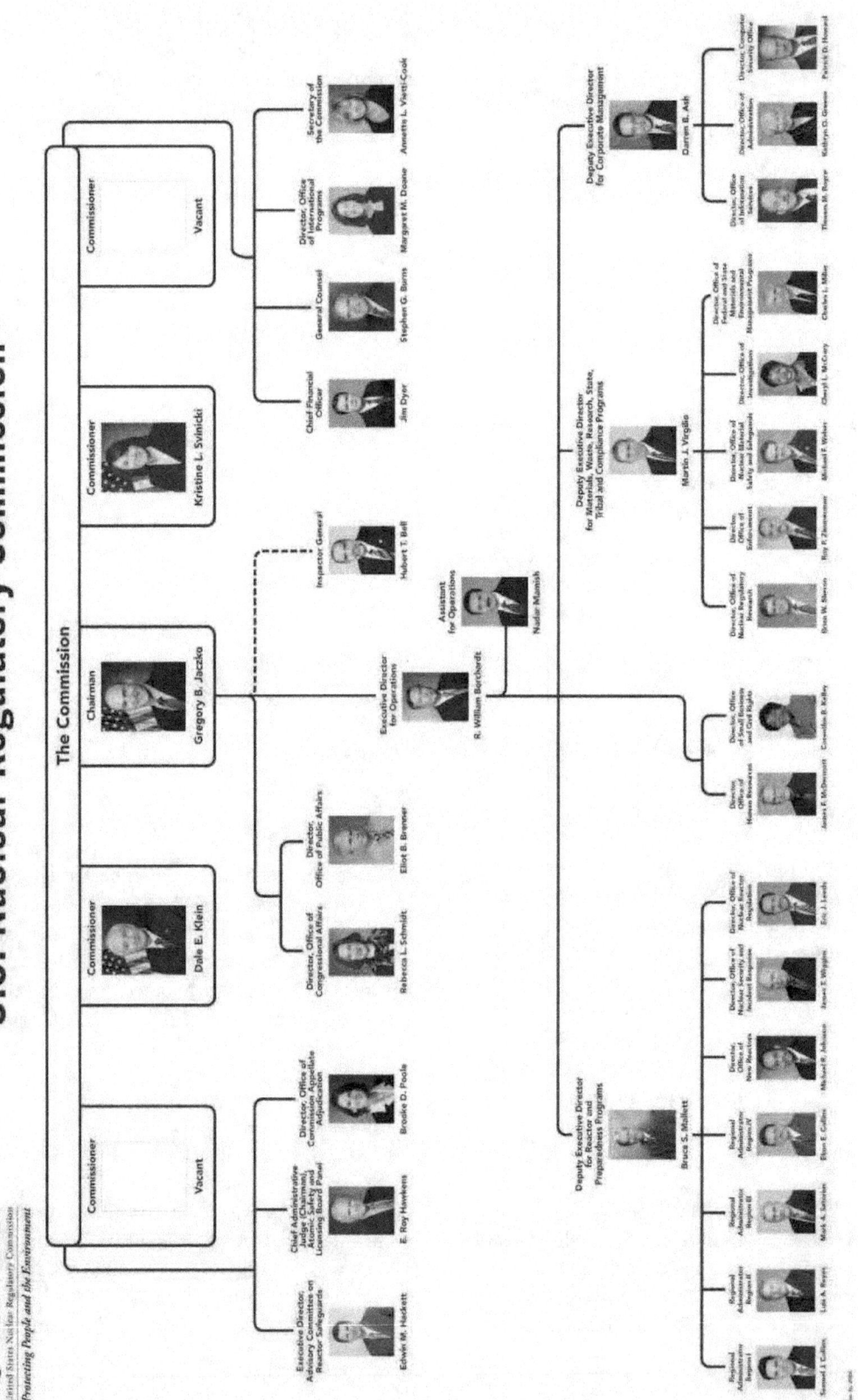

U.S. Nuclear Regulatory Commission

Figure 1
NRC BUDGETARY AUTHORITY, FY 2004–2009

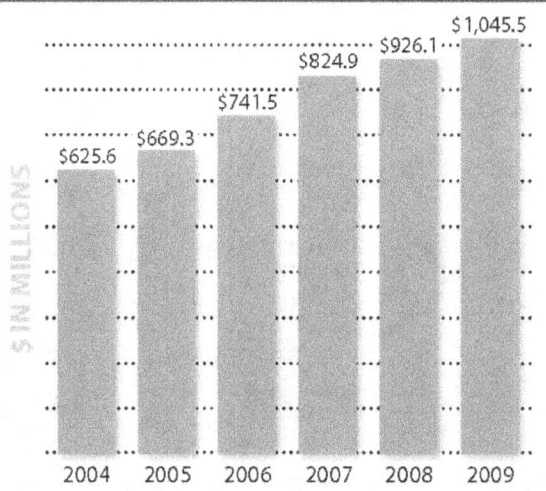

Source: NRC Performance Budget Fiscal Year 2010

Figure 2
NRC PERSONNEL CEILING, FY 2004–2009

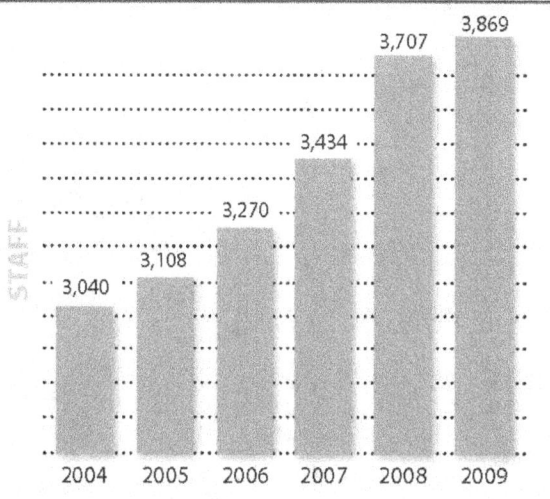

Source: NRC Performance Budget Fiscal Year 2010

is the focal point for the agency's communications with its licensees, State agencies, and other Federal agencies concerning operating events in the commercial nuclear sector. NRC operations officers staff the Operations Center 24 hours a day. Please refer to the NRC organizational chart on the previous page.

The NRC's budget for FY 2009 was $1,045.5 million (see Figure 1) with 3,869 full-time equivalent staff (see Figure 2). The NRC recovers approximately 90 percent of its appropriations from fees paid by NRC licensees.

The Nuclear Industry

The NRC regulates the commercial use of radioactive materials. The nuclear material cycle begins with the mining and production of nuclear fuel, continues with the use of nuclear fuel to power the Nation's 104 nuclear power plants, and ends with the safe transportation and storage of spent nuclear fuel and other nuclear waste. The NRC's regulatory programs ensure that radioactive materials are used safely and securely at every stage in the nuclear material cycle. Under the NRC's Agreement State program, 37 States have assumed primary regulatory responsibility over the industrial, medical, and other users of nuclear

materials in their States. The NRC works closely with these States to ensure that the States maintain public safety. To address safety and security issues, the NRC has developed regulatory practices, knowledge, and expertise specific to each activity in the nuclear material cycle.

Approximately 20 percent of the Nation's electricity is generated by the 104 NRC-licensed commercial nuclear reactors operating in 31 States (see Figure 3). The NRC oversees 3,000 licenses for medical, academic, industrial, and general uses of nuclear materials (see Figure 4). The agency conducts approximately 1,200 health and safety inspections of its nuclear materials licensees annually. In addition, the 37 Agreement States oversee 19,800 licensees.

The NRC, Agreement States, and their licensees share a common responsibility to protect public health and safety.

Fuel Facilities

The production of nuclear fuel begins at uranium mines where milled uranium ore is used to produce a uranium concentrate called "yellow cake." At a

Figure 3
U.S. COMMERCIAL NUCLEAR POWER REACTORS

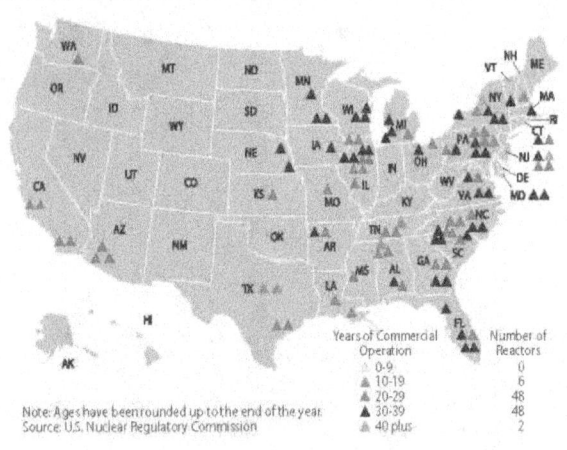

Years of Commercial Operation	Number of Reactors
0-9	0
10-19	6
20-29	48
30-39	48
40 plus	2

Note: Ages have been rounded up to the end of the year.
Source: U.S. Nuclear Regulatory Commission

Figure 4
U.S. MATERIALS LICENSEES

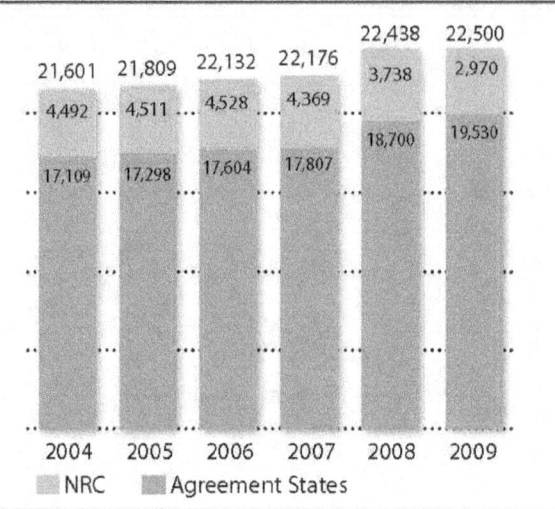

special facility, the yellow cake is converted into uranium hexafluoride gas and loaded into cylinders. The cylinders are sent to a gaseous diffusion plant, where uranium is enriched for use as reactor fuel. The enriched uranium is then converted into oxide powder, fabricated into fuel pellets (each about the size of a fingertip), loaded into metal fuel rods about 3.5 meters long, and bundled into reactor fuel assemblies at a fuel fabrication facility. Assemblies are then transported to nuclear power plants, nonpower research reactor facilities, and naval propulsion reactors for use as fuel. The NRC licenses eight major fuel fabrication and production facilities and three enrichment facilities in the United States. Because they handle extremely hazardous material, these facilities take special precautions to prevent theft, diversion by terrorists, and dangerous exposures to workers and the public from this nuclear material.

Reactors

Power plants change one form of energy into another. Electrical generating plants convert heat energy, the kinetic energy of wind or falling water, or solar energy into electricity. A nuclear power plant converts heat

energy into electricity. Other types of heat-conversion plants burn coal, oil, or gas to produce heat energy that is then used to produce electricity. Nuclear energy cannot be seen. There is no burning of fuel in the usual sense. Rather, energy is given off by the nuclear fuel as certain types of atoms split in a process called nuclear fission. This energy is in the form of fast-moving particles and invisible radiation. As the particles and radiation move through the fuel and surrounding water, the energy is converted into heat. The radiation energy can be hazardous, and facilities take special precautions to protect people and the environment from these hazards.

Because the fission reaction produces potentially hazardous radioactive materials, nuclear power plants are equipped with safety systems to protect workers, the public, and the environment. Radioactive materials require careful use because they produce radiation, a form of energy that can damage human cells. Depending on the amount and duration of the exposure, radiation can potentially cause cancer. In a nuclear reactor, most hazardous radioactive substances, called fission byproducts, are trapped in the fuel pellets or in the sealed metal tubes holding the fuel. However, small amounts of these radioactive fission byproducts, principally gases, become mixed with the water passing through the reactor. Other impurities in the water also become radioactive as they pass through the reactor. The facility processes and filters the water to remove these radioactive impurities and then returns the water to the reactor cooling system.

Materials Users

The medical, academic, and industrial fields all use nuclear materials. For example, about one-third of all patients admitted to U.S. hospitals are diagnosed or treated using radioisotopes. Most major hospitals have specific departments dedicated to nuclear medicine. In all, about 112 million nuclear medicine or radiation therapy procedures are performed annually, with the vast majority used in diagnoses. Radioactive materials used as a diagnostic tool can identify the status of a disease and minimize the need for surgery. Radioisotopes give doctors the ability to look inside the body and observe soft tissues and organs, in a manner similar to the way x rays provide images of bones. Radioisotopes carried in the blood also allow doctors to detect clogged arteries or check the functioning of the circulatory system.

Figure 5
SCHEMATIC OF THE NUCLEAR FUEL CYCLE

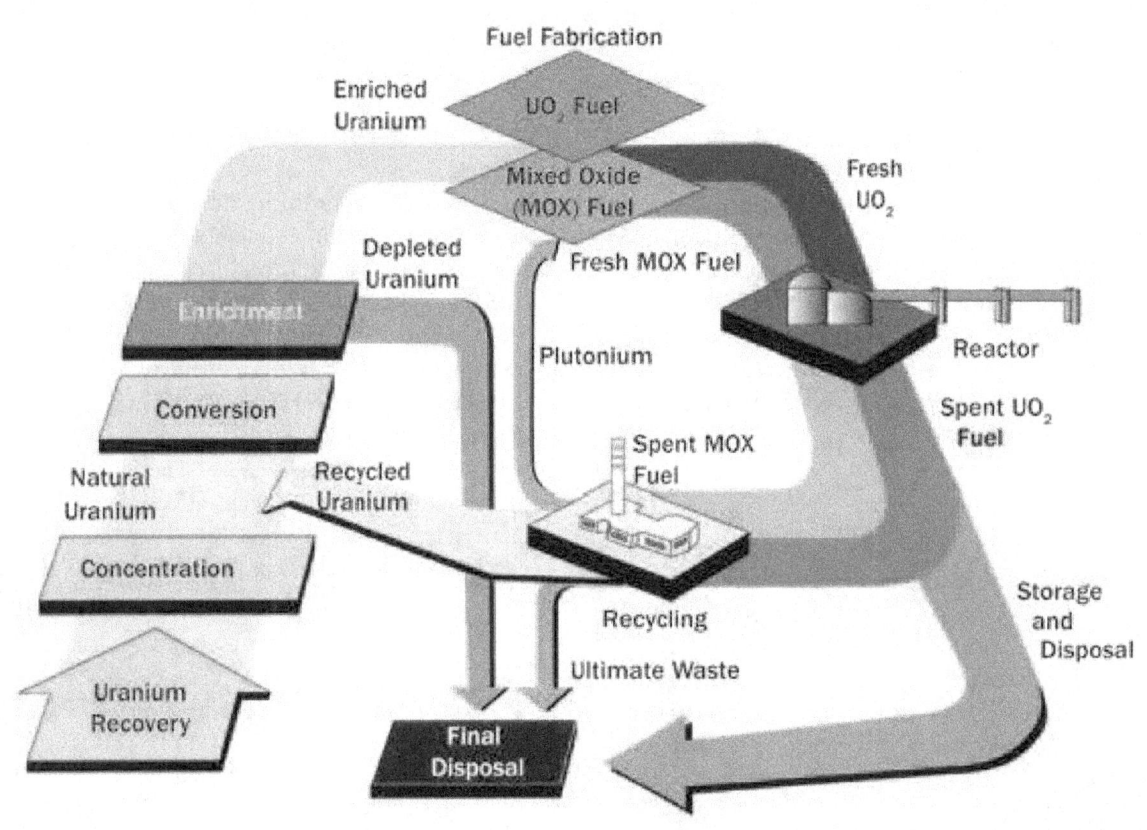

Source: U.S. Nuclear Regulatory Commission

The same property that makes radiation hazardous can also make it useful in treating certain diseases like cancer. When living tissue is exposed to high levels of radiation, cells can be destroyed or damaged. Doctors can selectively expose cancerous cells (cells that are dividing uncontrollably) to radiation to either destroy these cells or damage them so they can no longer reproduce.

Many of today's industrial processes also use nuclear materials. High-tech methods that ensure the quality of manufactured products often rely on radiation generated by radioisotopes. To determine whether a well drilled deep into the ground has the potential for producing oil, geologists use nuclear well-logging, a technique that employs radiation from a radioisotope inside the well to detect the presence of different materials. Radioisotopes are also used to sterilize instruments, find flaws in critical steel parts and welds that go into automobiles and modern buildings, authenticate valuable works of art, and solve crimes by spotting trace elements of poison. Radioisotopes can also eliminate dust from film and compact discs and reduce static electricity (which may create a fire hazard) from can labels. In manufacturing, radiation can change the characteristics of materials, often giving them features that are highly desirable. For example, wood and plastic composites treated with gamma radiation resist abrasion and require low maintenance. As a result, they are used for some flooring in high-traffic areas of department stores, airports, hotels, and churches.

Waste Disposal

During normal operations, a nuclear power plant generates the following two types of radioactive waste: high-level waste, which consists of used fuel (usually called spent fuel), and low-level waste, which includes contaminated equipment, filters, maintenance materials, and resins used in purifying water for the reactor cooling system. Other users of radioactive materials also generate low-level waste.

Nuclear power plants handle each type of radioactive waste differently. They must use special procedures in the handling of the spent fuel because it contains the highly radioactive fission byproducts created while the

reactor was operating. Typically, the spent fuel from nuclear power plants is stored in water-filled pools at each reactor site or at a storage facility in Illinois. The water in the spent fuel storage pool provides cooling and adequately shields and protects workers from the radiation. Several nuclear power plants have also begun using dry casks to store spent fuel. These heavy metal or concrete casks rest on concrete pads adjacent to the reactor facility. The thick layers of concrete and steel in these casks shield workers and the public from radiation.

Currently most spent fuel in the United States remains stored at individual plants. Permanent disposal of spent fuel from nuclear power plants requires a disposal facility that can provide reasonable assurance that the waste will remain isolated for thousands of years. The U.S. Department of Energy (DOE) submitted an application for a permanent, spent fuel disposal facility at Yucca Mountain, NV. This application is docketed and under review.

Licensees often store low-level waste onsite until its radioactivity has decayed, and the waste can be disposed of as ordinary trash, or until amounts are large enough for shipment to a low-level waste disposal site in containers approved by the U.S. Department of Transportation. The NRC has developed a waste classification system for low-level radioactive waste based on its potential hazards and has specified disposal and waste form requirements for each of the following general classes of waste: Class A, Class B, and Class C. Generally, Class A waste contains lower concentrations of radioactive material than Class B and Class C wastes. Two low-level disposal facilities accept a broad range of low-level wastes. They are located in Barnwell, SC, and Richland, WA.

Program Performance Overview

The NRC's FY 2008–2013 Strategic Plan determines the agency's long-term goals and strategic direction. The agency has two strategic goals: safety and security. To achieve its goals, the agency is organized into two major programs: the Nuclear Reactor Safety Program and the Nuclear Materials and Waste Safety Program.

FY 2009 Safety Goal

Performance Measures	2004	2005	2006	2007	2008	2009
Number of new conditions evaluated as red by the Reactor Oversight Process is ≤3.	1	0	0	0	0	0
Number of significant accident sequence precursors of a nuclear reactor accident is 0.	0	0	0	0	0	0
Number of operating reactors with integrated performance that entered the Manual Chapter 0350 process, the multiple/repetitive degraded cornerstone column, or the unacceptable performance column of the Reactor Oversight Process Action Matrix, with no performance exceeding Abnormal Occurrence Criterion I.D.4, is ≤4.	1	0	0	1	0	0
Number of significant adverse trends in industry safety performance, with no trend exceeding Abnormal Occurrence Criterion I.D.4, is ≤1.	0	0	0	0	0	0
Number of events with radiation exposures to the public and occupational workers that exceed Abnormal Occurrence Criterion I.A is:						
Reactors: 0	0	0	0	0	0	0
Materials: ≤3	0	1	0	0	0	0
Waste: 0	0	0	0	0	0	0
Number of radiological releases to the environment that exceed applicable regulatory limits is:						
Reactor: ≤3	0	0	0	0	0	0
Materials: ≤2	1	0	0	0	0	0
Waste: 0	0	0	0	0	0	0

FY 2009 Security Goal

Performance Measures	2004	2005	2006	2007	2008	2009
Number of unrecovered losses or thefts of risk-significant radioactive sources is 0.	0	0	0	0	0	0
Number of substantiated cases of theft or diversion of licensed, risk-significant radioactive sources or formula quantities of special nuclear material or number of attacks that result in radiological sabotage, is 0.	0	0	0	0	0	0
Number of substantiated losses of formula quantities of special nuclear material or substantiated inventory discrepancies of formula quantities of special nuclear material that are caused by theft or diversion or by substantial breakdown of the accountability system sabotage is 0.	0	0	0	0	0	0
Number of substantial breakdowns of physical security or material control that significantly weaken the protection against theft, diversion, or sabotage is <1.	0	0	0	0	0	0
Number of significant, unauthorized disclosures of classified and/or safeguards information is 0.	0	0	0	0	0	0

Nuclear Reactor Safety Program

The Nuclear Reactor Safety Program encompasses all NRC efforts to ensure that civilian nuclear power reactor facilities and research and test reactors are licensed and operated in a manner that adequately protects the public health and safety, preserves the environment, and protects against radiological sabotage and theft or diversion of special nuclear material.

Nuclear Materials and Waste Safety Program

The Nuclear Materials and Waste Safety Program focuses on the safe and secure use of remaining radioactive materials. The Nuclear Materials and Waste Safety Program regulates fuel facilities, medical and industrial nuclear materials users, the disposal of both high-level and low-level waste, the decommissioning of power plants, and the storage and transportation of spent nuclear fuel.

NRC PERFORMANCE MEASURE RESULTS

STRATEGIC GOALS	
Safety	Security

MAJOR PROGRAMS	
Nuclear Reactor Safety	Nuclear Materials and Waste Safety

ACTIVITIES	
Nuclear Reactor Licensing	Fuel Facilities
	High-Level Waste Repository
Nuclear Reactor Inspection	Nuclear Materials Users
	Decommissioning and Low-Level Waste
	Spent Fuel Storage and Transportation

Program Performance Results

STRATEGIC GOAL 1: SAFETY
Ensure Adequate Protection of Public Health and Safety and the Environment

Safety is the primary goal of the NRC. The agency achieves this goal by ensuring that the performance of licensees is at or above acceptable safety levels. NRC safety programs work in conjunction with our licensees in a partnership. The NRC licensees are responsible for designing, constructing, and operating nuclear facilities safely. The NRC is responsible for regulatory oversight of the licensees. The NRC designed its safety goal activities to achieve the following strategic outcomes:

Strategic Outcomes

- Prevent the occurrence of any nuclear reactor accidents.
- Prevent the occurrence of any inadvertent criticality events.
- Prevent the occurrence of any acute radiation exposures resulting in fatalities.
- Prevent the occurrence of any releases of radioactive materials that result in significant radiation exposures.
- Prevent the occurrence of any releases of radioactive materials that cause significant adverse environmental impacts.

FY 2009 Results

In FY 2009, the NRC achieved all five of its safety goal strategic outcomes. The NRC also uses six performance measures to determine whether it has met its safety goal. The agency met all six performance measure targets in FY 2009.

Three of the performance measures focus on performance at individual nuclear power plants. Inspection results show that all of the nuclear power plants are operating safely. The fourth measure tracks the trends of several key indicators of nuclear power plant safety. This measure is the broadest measure of the safety of nuclear power plants, incorporating the performance results from all plants to determine industry average results. The measure results show that there were no statistically significant adverse trends in any of the indicators in FY 2009.

The last two safety performance measures track harmful radiation exposures to the public and occupational workers and radiation exposures that harm the environment. None of these measures exceeded their targets in FY 2009.

STRATEGIC GOAL 2: SECURITY

Ensure Adequate Protection in the Secure Use and Management of Radioactive Materials

The NRC must remain vigilant in ensuring the security of nuclear facilities and materials in an elevated threat environment. The agency achieves its common defense and security goal using licensing and oversight programs similar to those employed in achieving its safety goal. The NRC has designed its strategic goal activities to achieve the following strategic outcome:

Strategic Outcome

- Prevent any instances in which licensed radioactive materials are used domestically in a manner hostile to the security of the United States.

FY 2009 Results

In FY 2009, the NRC achieved its security goal strategic outcome. The NRC also uses five security goal performance measures to determine whether the agency has met its security goal. The agency met all five performance measure targets in FY 2009. The first performance measure tracks unrecovered losses or thefts of risk-significant radioactive sources. The measure ensures that those radioactive sources that the agency has determined to be risk-significant to the public health and safety are accounted for at all times. The ability to account for these sources is critical to

secure the nation from "dirty bomb" attacks or other means of radiation dispersal.

The second, third, and fourth performance measures evaluate the number of significant security events and incidents that occur at NRC-licensed facilities. These measures determine whether nuclear facilities maintain adequate protective forces to prevent theft or diversion of nuclear material or sabotage; whether systems in place at licensee plants accurately account for the type and amount of materials processed, utilized, or stored; and whether the facilities account for special nuclear material at all times with no losses of this material. No events met the conditions for this measure in FY 2009.

The last security measure tracks significant unauthorized disclosures of classified or safeguards information that may cause damage to national security or public safety. This measure focuses on whether classified information or safeguards information is stored and utilized in such a way as to prevent its disclosure to the public, terrorist organizations, other Nations, or personnel without a need to know. Unauthorized disclosures can harm national security or compromise public health and safety. The measure also focuses on whether controls are in place to maintain and secure the various devices and systems (electronic or paper based) which the agency and its licensees use to store, transmit, and utilize this information. No documented disclosures of this type of information occurred during FY 2009.

Data Completeness and Reliability

The NRC considers the data contained in this report to be complete, reliable, and relevant. The data are complete because the agency reports actual performance data for every performance goal and indicator in the report. The agency also considers the data in this report reliable and relevant because the data have been validated and verified. "Verification and Validation of NRC's Performance Measures," contains the processes the agency uses to collect, validate, and verify performance data in this report. Please visit http://www.nrc.gov/about-nrc/fy2009-par-verification.pdf to view this report in its entirety.

Future Challenges

The NRC ensures that the health and safety of the American public and the environment are adequately protected from any harmful effects of using nuclear materials. The industry has experienced a substantial improvement in safety at nuclear power plants over the past 20 years as both the nuclear industry and the NRC have gained substantial experience in the operation and maintenance of nuclear power facilities. However, despite the excellent safety and security record of the industry, the agency cannot rest on its achievements. The primary challenges the agency faces are the large number of new nuclear plants that have applied for licenses, the safe disposal of high-level nuclear waste, and the need to ensure security at nuclear facilities.

New Nuclear Power Plants

With increased concerns about the continued availability and cost of oil, as well as concerns over the environmental damage caused by coal-burning electrical plants, the amount of electricity supplied by nuclear power is likely to increase substantially in the future. The NRC last issued a nuclear power plant construction permit in 1977. To date, the agency has docketed a total of 18 combined operating license (COL) applications for sites across the country. The agency's primary challenge is to license new reactors to ensure that they will operate safely as they provide electricity required by the Nation for economic growth. However, before licensing any new nuclear reactor, the agency requires a detailed analysis of new reactor designs. This analysis includes a study of the reactor's vulnerability to accidents and security compromises. It also includes the development of inspection procedures, tests, analyses, and acceptable criteria for construction. The NRC is also evaluating commercial gas centrifuge facilities that utilize new methods of enriching nuclear fuel for reactors.

Safe Disposal of High-Level Waste

Safely disposing of the waste from nuclear power plants is vital to protecting public health and the environment. In FY 2008, DOE filed a license application to establish the Nation's first repository for high-level radioactive waste at Yucca Mountain,

NV. The NRC accepted and docketed the application. The agency has begun a review to evaluate a wide range of technical and scientific issues and will attempt to resolve regulatory concerns. Most nuclear waste is now safely and securely stored at reactor sites. In addition to the storage of nuclear waste, safely transporting spent nuclear fuel is a significant issue for the public and the agency. More than 1,300 spent fuel shipments regulated by the NRC have been safely transported in the United States in the past 25 years. Therefore, the agency must be able to assure the public that all movements of nuclear waste, including those to a permanent storage site, will be safe and secure.

Security at Nuclear Facilities

In addition to safety, the security of nuclear materials is of paramount importance to the Nation. Nuclear facilities are among the most secure facilities in the Nation. The NRC, in concert with other Federal agencies, constantly monitors intelligence to determine the level of threat faced by nuclear facilities. The agency continues to improve the regulatory requirements to better ensure the security of nuclear materials and facilities. The threat faced by the Nation from those seeking to steal classified information has become more urgent in recent years. Nuclear facilities have implemented increased security measures, including "force-on-force" training exercises, to help ensure protection of this vital national infrastructure.

Financial Performance Overview

As of September 30, 2009, the financial condition of the NRC was sound with respect to having sufficient funds to meet program needs and adequate control of these funds in place to ensure obligations did not exceed budget authority. The NRC prepared its financial statements in accordance with the accounting standards codified in the Statements of Federal Financial Accounting Standards (SFFAS) and Office of Management and Budget (OMB) Circular A-136, "Financial Reporting Requirements."

Sources of Funds

The NRC has two appropriations, Salaries and Expenses and Office of the Inspector General. Funds for both appropriations are available until expended. The NRC's total new FY 2009 budget authority was $1,045.5 million. Of this amount, $1,034.6 million was for the Salaries and Expenses appropriation and $10.9 million was for the Office of the Inspector General appropriation. This represents an increase in new budget authority of $119.4 million over FY 2008 ($117.3 million for the Salaries and Expenses appropriation and $2.1 million for the Office of the Inspector General appropriation). In addition, $100.0 million from prior-year appropriations, $7.4 million from prior-year reimbursable work, and $12.3 million for new reimbursable work to be performed for others was available to obligate in FY 2009. The sum of all funds available to obligate for FY 2009 was $1,165.2 million, which was a $136.4 million increase over the FY 2008 amount of $1,028.8 million.

Figure 6
SOURCES OF FUNDS

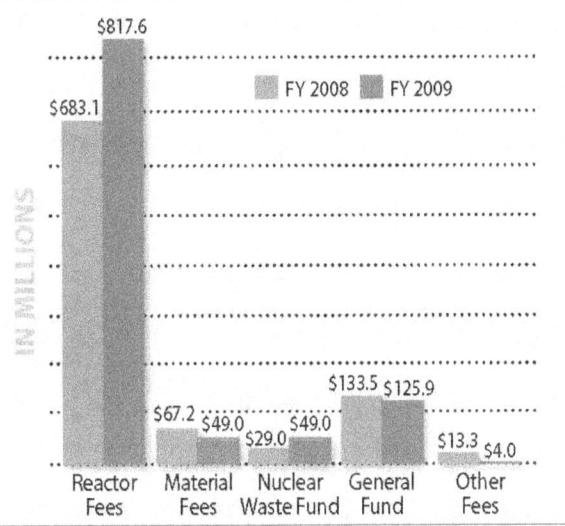

The Omnibus Budget Reconciliation Act of 1990 (OBRA-90), as amended, required the NRC to collect fees to offset approximately 90 percent of its new budget authority, less the amount appropriated to

the NRC from the Nuclear Waste Fund and amounts appropriated for waste incidental to reprocessing and generic homeland security for FY 2009. The NRC collected $857.8 million in reactor and material fees in FY 2009. This is 98.5 percent of the fee recovery requirement.

Uses of Funds by Function

The NRC incurred obligations of $1,084.1 million in FY 2009, which was an increase of $134.3 million over FY 2008. Approximately 53 percent of obligations were used for salaries and benefits. The remaining 47 percent was used to obtain technical assistance for the NRC's principal regulatory programs, to conduct confirmatory safety research, to cover operating expenses (e.g., building rentals, transportation, printing, security services, supplies, office automation, training), staff travel, and reimbursable work. The unobligated budget authority available at the end of FY 2009 was $81.1 million, an increase compared to the FY 2008 amount of $79.0 million. Of this $81.1 million, $9.3 million was for reimbursable work and $71.8 million was available to fund critical NRC needs in FY 2009.

Figure 7
USES OF FUNDS BY FUNCTION

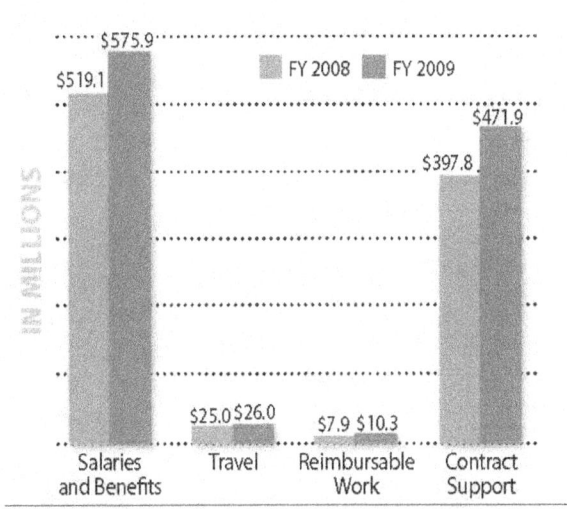

Audit Results

The NRC received an unqualified audit opinion on its FY 2009 financial statements. In FY 2008, the auditors identified a significant deficiency related to the method by which the NRC estimates the accounts payable balance which represents costs for billed and unbilled goods and services received (prior to year end) that are unpaid. Prior to the last quarter of FY 2008, the NRC used an algorithm that recognized accounts payable as a specific percentage of NRC's total expenses to date. Once this percentage was calculated, it was applied to an annualized expense figure. In the fourth quarter of FY 2008, the NRC implemented a revised methodology to calculate the accounts payable estimate. The new methodology involves analyzing the actual activity for the largest obligations to include in the estimate. For the remaining smaller obligations, the agency analyzed actual activity of a percentage of the obligations and developed an algorithm to estimate the total amount to include in the accounts payable balance. Throughout FY 2009, the NRC continued to refine this methodology and validated the estimate each quarter. In FY 2009, the auditors closed this significant deficiency due to the accuracy of this new estimation methodology.

In FY 2007 and FY 2008, the auditors also identified the Fee Billing System as a substantial noncompliance with the Federal Financial Management Improvement Act (Improvement Act) because of a lack of current certification and accreditation. In FY 2009, the auditors closed the substantial noncompliance because the NRC completed the certification and accreditation of the Fee Billing System.

A summary of the Financial Statement Audit Results is included in the "Other Accompanying Information" section of this report.

Limitations of the Financial Statements

The principal statements have been prepared to report the financial position and results of operations of the NRC, pursuant to the requirements of 31 U.S.C. 3515 (b). While the statements have been prepared from the books and records of the NRC in accordance with generally accepted accounting principles (GAAP) for Federal entities and the formats prescribed by the OMB, the statements are in addition to the financial reports used to monitor and control budgetary resources, which are prepared from the same books

and records. The statements should be read with the realization that they are for a component of the U.S. Government, a sovereign entity.

Financial Statement Highlights

The NRC's financial statements summarize the financial activity and financial position of the agency. Chapter 3 presents the financial statements, footnotes, and required supplementary information. Analysis of the principal statements follows.

Analysis of the Balance Sheet

The NRC's assets were $611.8 million as of September 30, 2009, an increase of $57.3 million from the end of FY 2008. The increase is due to the Fund Balance with the U.S. Department of the Treasury (Treasury) increasing by $55.1 million as a result of an increase in appropriated funds received which were obligated but not yet disbursed.

ASSET SUMMARY (In Millions)

As of September 30,	2009	2008
Fund Balance with Treasury	$448.6	$393.5
Accounts Receivable, Net	128.2	121.4
Property & Equipment, Net	31.6	35.5
Other	3.4	4.1
Total Assets	$611.8	$554.5

The Fund Balance with Treasury was $448.6 million at September 30, 2009, accounting for 73 percent of total assets. This account represents appropriated funds, collected license fees, and other funds maintained at Treasury to pay current liabilities and to finance authorized purchase commitments. The $55.1 million increase in the balance reflects an increase of $119.4 million in new budget authority (including an increase of $20.0 million for the Nuclear Waste Fund transfer), a $37.1 million beginning balance increase over the prior year beginning balance, and an increase of $107.4 million in fee collections, less expenditure increases of $60.1 million in salaries and benefits, $46.6 million in general disbursements, $8.5 million in grant disbursements, and an increase in fee collection transfers to Treasury of $94.2 million. The difference between the increase in fee collections and fee collections transferred to Treasury results from an over collection of $13.3 million in fees during FY 2007, which were included in the FY 2008 fee transfer to Treasury.

Accounts receivable consists of amounts owed to the NRC by other Federal agencies and the public. Accounts Receivable, Net as of September 30, 2009, was $128.2 million, which includes an offsetting allowance for doubtful accounts of $3.1 million. This 6 percent increase from the FY 2008 year-end Accounts Receivable, Net balance of $121.4 million is the result of an increase of $11.4 million in licensing and inspection activities due to an increase in hours invoiced and in the hourly rate for the NRC's services, offset by a decrease of $5.4 million in accruals for materials and facilities open inspections.

LIABILITIES SUMMARY (In Millions)

As of September 30,	2009	2008
Accounts Payable	$ 51.0	$ 54.1
Federal Employee Benefits	7.6	7.1
Other Liabilities	86.2	75.8
Total Liabilities	$144.8	$137.0

Total liabilities were $144.8 million as of September 30, 2009, an increase of $7.8 million from the FY 2008 year-end balance of $137.0 million. The increase resulted from an increase in Other Liabilities of $10.4 million, which comprises increases of $3.6 million in accrued annual leave, $3.5 million in accrued funded salaries and benefits, and $2.4 million in grants payable.

Of the agency's liabilities, $56.6 million was not covered by budgetary resources, an 8 percent increase over the balance of $52.5 million as of September 30, 2008. The increase of $4.1 million was primarily due to an increase in unfunded accrued annual leave of $3.6 million resulting from an increase in the number of full-time employee equivalents and salary increases. The liabilities not covered by budgetary resources in FY 2009 include $47.3 million in unfunded accrued annual leave included in Other Liabilities, for the amount of leave, earned but not yet taken and $7.6 million in future workers' compensation included in Federal Employee Benefits.

NET POSITION SUMMARY (In Millions)

For the Years Ended September 30,	2009	2008
Unexpended Appropriations	$338.6	$289.3
Cumulative Results of Operations	128.4	128.2
Total Net Position	$467.0	$417.5

Net Position, the difference between Total Assets and Total Liabilities, was $467.0 million as of September 30, 2009, an increase of $49.5 million from the FY 2008 year-end balance. Net Position is comprised of two components: Unexpended Appropriations and Cumulative Results of Operations. Unexpended Appropriations is the amount of spending authority granted by Congress that remains unused by the agency. The increase in FY 2009 for Unexpended Appropriations is $49.3 million. Cumulative Results of Operations, which represents the cumulative excess of financing sources over expenses, remained relatively constant at September 30, 2009 and 2008.

Analysis of the Statement of Net Cost

Net costs are gross costs offset by earned revenue. The Statement of Net Cost presents the net cost of NRC's two programs as identified in the NRC Annual Performance Plan. The purpose of this statement is to link program performance to the cost of programs. The NRC's Net Cost of Operations for the year ended September 30, 2009, was $170.4 million, which is an increase of $23.9 million over the FY 2008 net cost of $146.5 million.

NET COST OF OPERATIONS (In Millions)

For the Years Ended September 30,	2009	2008
Nuclear Reactor Safety and Security	$ 2.9	$ (20.0)
Nuclear Materials and Waste Safety and Security	167.5	166.5
Net Cost of Operations	$ 170.4	$ 146.5

NRC's total gross costs increased $98.8 million. The Nuclear Reactor Safety and Security program gross costs increased $91.1 million primarily because of increases of $51.2 million in salaries and benefits, $24.3 million in contractor support, and $7.8 million in grants for nuclear education. These increases are primarily in the areas of new reactor activities, and existing licensing and oversight activities. The Nuclear Materials and Waste Safety and Security program gross costs increased $7.7 million primarily in the areas of nuclear materials licenses, fuel facilities, and decommissioning activities.

Total earned revenue increased $74.9 million from $797.6 million for the year ended September 30, 2008, to $872.5 million at September 30, 2009. Earned revenue increased for the Nuclear Reactor Safety and Security program by $68.2 million and for the Nuclear Materials and Waste Safety and Security program by $6.7 million. The increases primarily result from increases in fees collected due to the increase in appropriations for NRC activities, of which the NRC is required to collect approximately 90 percent through fee billing. Fees for reactor and materials licensing and inspections are collected in accordance with Title 10 of the *Code of Federal Regulations* (10 CFR) Part 170, "Fees for Facilities, Materials, Import and Export Licenses, and Other Regulatory Services under the Atomic Energy Act of 1954, as Amended," and 10 CFR Part 171, "Annual Fees for Reactor Licenses and Fuel Cycle Licenses and Materials Licenses, Including Holders of Certificates of Compliance, Registrations, and Quality Assurance Program Approvals and Government Agencies Licensed by the NRC."

Analysis of the Statement of Changes in Net Position

The Statement of Changes in Net Position reports the change in net position during the reporting period. Net position is affected by changes in its two components—Cumulative Results of Operations and Unexpended Appropriations. The increase in Net Position of $49.5 million from FY 2008 to FY 2009, was the result of an increase in Unexpended Appropriations. A change in unexpended appropriations results from appropriations received being more, or less, than appropriations used during the fiscal year. In FY 2009, appropriations received of $138.7 million consisted of NRC's total appropriation of $1,045.5 million, reduced by $857.8 million in fee collections returned to Treasury and the Nuclear Waste Fund transfer of $49.0 million. Appropriations used in FY 2009 totaled $89.3 million and consisted of funds used of $993.9 million reduced by collection from fees assessed of $857.8 million and Nuclear Waste Fund expenses of $46.8 million.

Analysis of the Statement of Budgetary Resources

The Statement of Budgetary Resources reports the source and status of budgetary resources at the end of the period. It presents the relationship between budget authority and budget outlays and the reconciliation of obligations to total outlays. For FY 2009, NRC had total budgetary resources available of $1,165.2 million, a 13 percent increase over FY 2008 budgetary resources available of $1,028.8 million. The increase primarily resulted from an increase in appropriations received of $119.4 million which increased from $926.1 million in FY 2008 to $1,045.5 million in FY 2009. The appropriation included increases of $47.7 million for the Nuclear Reactor Safety and Security program, $69.6 million for the Nuclear Materials and Waste Safety and Security program, and $2.1 million for the Office of the Inspector General. This funding provided for increases in salaries and benefits of $49.0 million and contract support services of $70.4 million, primarily for growth of regulatory and support activities for new reactor facilities, regulatory oversight of existing reactors, and existing materials and waste facilities licensing activities.

For FY 2009, the NRC had Obligations Incurred of $1,084.1 million, compared to FY 2008 Obligations Incurred of $949.8 million, an increase of $134.3 million. The increase resulted primarily from an increase in salaries and benefits and contract support (see Figure 7). Gross outlays for FY 2009 were $999.1 million, which represents an increase of $115.1 million over FY 2008 gross outlays of $884.0 million. The increase resulted from an increase in salaries and benefits disbursements of $60.1 million, general disbursements of $46.6 million, and grant disbursements of $8.5 million. Gross outlay increases are reflected in the Nuclear Reactor Safety and Security program at $96.0 million, primarily for new reactor and existing reactor licensing activities; the Nuclear Materials and Waste Safety and Security program at $9.2 million, primarily for materials licensing, fuel facilities, and decommissioning; and nuclear education grants of $8.5 million.

Systems, Controls, and Legal Compliance

Management Assurances

This section provides information on the NRC's compliance with the Federal Managers' Financial Integrity Act, the OMB Circular A-123, "Management's Responsibility for Internal Control," and the Federal Financial Management Improvement Act. Other Accompanying Information" section the, "Summary of Financial Statement Audit and Management Assurances," includes a summary of management assurances.

Federal Managers Financial Integrity Act

The Integrity Act mandates that agencies establish controls to reasonably ensure that the agency (1) complies with applicable laws concerning obligations and costs; (2) safeguards assets against waste, loss, unauthorized use, or misappropriation; and (3) properly accounts for and records revenues and expenditures. The Integrity Act encompasses program, operational, and administrative areas, as well as accounting and financial management. It also requires the Chairman to provide an assurance statement on the adequacy of internal controls and on the conformance of financial systems with Governmentwide standards.

Management Control Review Program

Managers throughout the NRC are responsible for implementing effective controls in their areas of responsibility. Each office director and regional administrator prepares an annual assurance certification that identifies any control weaknesses requiring the attention of the NRC's Executive Committee on Internal Control (ECIC). These statements are based on various sources, including management knowledge gained from the daily operation of agency programs, management reviews, program evaluations, audits of financial statements, reviews of financial systems, annual performance

**U.S. NUCLEAR REGULATORY COMMISSION
FEDERAL MANAGERS' FINANCIAL INTEGRITY ACT
STATEMENT FOR FY 2009**

The U.S. Nuclear Regulatory Commission's (NRC) management is responsible for establishing and maintaining effective internal control and financial management systems that meet the objectives of the Federal Managers' Financial Integrity Act (Integrity Act). The NRC conducted its assessment of internal control over the effectiveness and efficiency of operations and compliance with applicable laws and regulations, and in accordance with OMB Circular A-123, *Management's Responsibility for Internal Control*. Based on the results of this evaluation, the NRC can provide reasonable assurance that its internal control over the effectiveness and efficiency of operations and compliance with applicable laws and regulations as of September 30, 2009, was operating effectively and no material weaknesses were found in the design or operation of internal control.

NRC can also provide reasonable assurance that its financial systems substantially conform to the Integrity Act and comply with the component requirements of the Federal Financial Management Improvement Act.

In addition, the NRC conducted its assessment of the effectiveness of internal control over financial reporting, which includes safeguarding of assets and compliance with applicable laws and regulations, in accordance with the requirements of Appendix A of OMB Circular A-123. Based on the results of the evaluation, the NRC can provide reasonable assurance that NRC's internal control over financial reporting as of June 30, 2009, was operating effectively, and no material weaknesses were found in the design or operation of the internal control over financial reporting.

Gregory B. Jaczko
Chairman
U.S. Nuclear Regulatory Commission
November 13, 2009

plans, Inspector General and U.S. Government Accountability Office reports, and reports and other information provided by the congressional committees of jurisdiction.

The NRC's ECIC includes senior executives from the Office of the Chief Financial Officer and the Office of the Executive Director for Operations. Staff from the Office of the General Counsel participates as an advisor.

The ECIC met and reviewed the assurance certifications provided by the offices and regions. The ECIC then informed the Chairman as to whether the NRC had any internal control deficiencies serious enough to require reporting as a material weakness or noncompliance.

The NRC's ongoing internal control program requires, among other things, that reports on internal control deficiencies be integrated into the offices' and regions' annual operating plans. The operating plan process provides for periodic updates and ensures that key issues receive senior management attention. Combined with the individual assurance statements discussed previously, the internal control information in these plans provides the framework for monitoring and improving the agency's internal controls on an ongoing basis.

FY 2009 Integrity Act Results

The NRC evaluated its internal control systems for the fiscal year ending September 30, 2009. Based on this evaluation, the NRC is able to provide a statement of assurance that the internal controls and financial management systems meet the objectives of the Integrity Act. The NRC has reasonable assurance that its internal controls are effective and that its financial management systems conform to Governmentwide standards.

Office of Management and Budget Circular A-123, "Management's Responsibility for Internal Control," including Appendix A, "Internal Control over Financial Reporting"

In FY 2006, the NRC implemented the requirements of the OMB revised Circular A-123, which defined and strengthened management's responsibility for internal control in Federal agencies. The revised circular included updated internal control standards. A new section, Appendix A, required Federal agencies to assess the effectiveness of internal controls over their financial reporting and to prepare a separate annual statement of assurance as of June 30, 2009.

In FY 2007, the agency adopted a 3-year rotational testing plan. The NRC determined that three of the original nine key processes were significant enough to include in the testing each year of the 3-year cycle. The remaining six key processes were to be tested once in the 3-year cycle, two each year. In FY 2008 and 2009, the NRC continued its assessment of internal control over financial reporting. The agency reevaluated its scope of financial reports, materiality values, risk assessments, key processes, and key controls. Based on the results of this evaluation, the NRC can provide reasonable assurance that its internal control over financial reporting was operating effectively as of June 30, 2009, and that the evaluation found no material weaknesses in the design or operation of the internal controls over financial reporting.

Federal Financial Management Improvement Act

The Federal Financial Management Improvement Act (Improvement Act) requires each agency to implement and maintain systems that comply substantially with (1) Federal financial management system requirements, (2) applicable Federal accounting standards, and (3) the standard general ledger at the transaction level. The Improvement Act requires the Chairman to determine whether the agency's financial management systems comply with the Improvement Act and to develop remediation plans for systems that do not comply.

FY 2009 Improvement Act Results

As of September 30, 2009, the NRC evaluated its financial systems to determine if they complied with applicable Federal requirements and accounting standards required by the Improvement Act. The NRC evaluated the following eight systems: the Federal Financial System, Federal Personnel Payroll System, Human Resources Management System, Cost Accounting System, Advice of Allotments/Financial Plan System, Capitalized Property System, Fee Billing System, and Controller Resource Database System. As of September 30, 2009, the agency's financial management systems were in compliance with the Improvement Act. In making this determination, the NRC considered all the information available, including the report from the ECIC on the effectiveness of internal controls, the Office of the Inspector General audit reports, and the results of the agency's financial management system reviews. The agency also relied on the Department of the Interior National Business Center (DOI-NBC) annual reasonable assurance statement, which concluded that, for FY 2009, the cross-serviced financial systems were in substantial compliance with Federal financial management system requirements.

In FY 2008, the financial management systems were in compliance with the Improvement Act, except for the Licensee Fee Billing System (Fee System) which was operating without its accreditation and Authority to Operate (ATO). The ATO was granted June 2009, therefore, the Inspector General closed the finding.

Figure 8
PROMPT PAYMENT

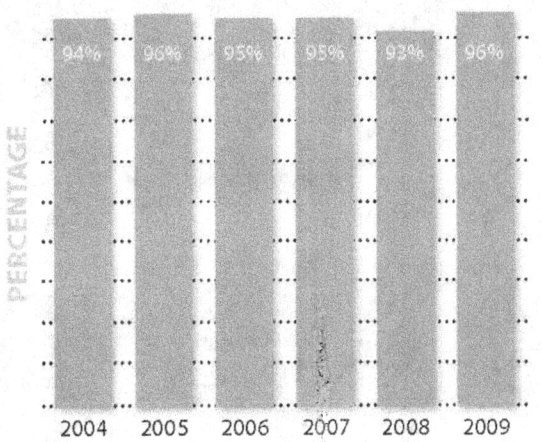

Prompt Payment

The Prompt Payment Act requires Federal agencies to make timely payments to vendors for supplies and services, to pay interest penalties when payments are made after the due date, and to take cash discounts when they are economically justified. In FY 2009, the NRC paid 96 percent of the 12,903 invoices subject to the

Prompt Payment Act on time (see Figure 8). The NRC incurred $19,825 in interest penalties during FY 2009.

Improper Payments

The NRC remains at low risk of making improper payments. At the present time, the NRC's payments consist of commercial vendor, interagency, and travel reimbursements. The NRC monitors and reports improper payments within its programs and continues to evaluate internal controls guarding against improper payments. The NRC continues to perform annual risk assessments for each of these areas. Based on the FY 2009 risk assessments, the number and amount of improper payments fall below the external reporting requirement established by OMB guidance on what is considered a significant risk. The NRC awards less than $500 million in annual contracts and, therefore, is not subject to annual reporting under the Recovery Auditing Act. The DOI-NBC's Federal Personnel/Payroll System, as the system of record for payroll disbursements, is responsible for monitoring and reporting on any improper payroll-related payments.

Figure 9
DELINQUENT DEBT

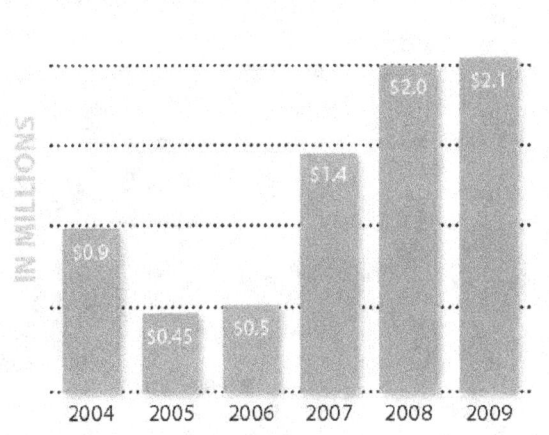

Debt Collection

The Debt Collection Improvement Act enhances the ability of the Federal Government to service and collect debts. The agency's goal is to maintain the level of delinquent debt owed to the NRC at year end to less than 1 percent of its annual billings. The NRC continues to meet this goal and, at the end of FY 2009, delinquent debt was $2.1 million (Figure 9). The NRC continues to pursue the collection of delinquent debt and refers all eligible debt over 180 days delinquent to the Treasury for collection.

Biennial Review of User Fees

The Chief Financial Officers Act requires agencies to conduct a biennial review of fees, royalties, rents, and other charges imposed by agencies and to make revisions to cover program and administrative costs incurred. Each year, the NRC revises the hourly rates for license and inspection fees and adjusts the annual fees to meet the fee collection requirements of OBRA-90, as amended. The *Federal Register* (74 FR 27641, June 10, 2009) describes the most recent changes to the

license, inspection, and annual fees. In FY 2009, the NRC revised the fees for public use of the auditorium to more appropriately recover the actual costs. The agency concluded that other types of fees did not warrant revisions at this time.

Inspector General Act

The agency has established and continues to maintain an excellent record in resolving and implementing Office of the Inspector General open audit recommendations. In the Other Accompanying Information section of this report, "Management Decisions and Final Actions on OIG Audit Recommendations," includes this information, as well as data concerning disallowed costs determined through contract audits conducted by the Defense Contract Audit Agency.

Nuclear energy fuel rods

Photo Courtesy of NRC Photo Library

Chapter 2

Program Performance

Turkey Point nuclear power plant is located on Biscayne Bay, south of Miami, FL, and just east of the Homestead area. It is run by Florida Power & Light Company.

Photo Courtesy of NRC Photo Library

Senior Construction Inspector Rashean Jackson reviews drawings during a problem identification and resolution inspection at the Louisiana Energy Services National Enrichment Facility (LES NEF). The LES NEF is gas centrifuge uranium enrichment facility that is currently under construction.

Measuring and Reporting Performance

This chapter presents information on the U.S. Nuclear Regulatory Commission's (NRC's) performance in achieving its mission during fiscal year (FY) 2009. The agency's mission is to license and regulate the Nation's civilian use of byproduct, source, and special nuclear materials to ensure adequate protection of public health and safety, promote the common defense and security, and protect the environment.

This chapter describes the NRC's performance results and program achievements in accomplishing its two strategic goals of safety and security. The safety goal discussion addresses the NRC's key regulatory oversight for operating reactor licensing, new reactor licensing, reactor inspection, fuel facilities, nuclear material users, high-level waste repository, decommissioning and low-level waste, and spent fuel storage and transportation. The security goal discussion addresses security activities in the nuclear reactor safety and nuclear materials and waste safety programs. In addition, this chapter describes the agency's progress in achieving its Organizational Excellence Objectives of openness, effectiveness, and operational excellence. Lastly, it describes information on data sources, data quality, and the completeness and reliability of performance data. The discussion focuses primarily on the NRC's methods for collecting and analyzing data, ensuring data security, and improving the agency's performance measures and the quality of its data during the current reporting period.

Goals and Performance Measures

STRATEGIC GOAL 1: SAFETY

Ensure Adequate Protection of Public Health and Safety and the Environment

Strategic Outcomes

The NRC uses the following five strategic outcomes associated with the safety goal that determine whether the agency has achieved its objective to ensure adequate protection of public health and safety and the environment:

- Prevent the occurrence of any nuclear reactor accidents.
- Prevent the occurrence of any inadvertent criticality events.
- Prevent the occurrence of any acute radiation exposures resulting in fatalities.
- Prevent the occurrence of any releases of radioactive materials that result in significant radiation exposures.
- Prevent the occurrence of any releases of radioactive materials that cause significant adverse environmental impacts.

RESULTS: In FY 2009, the NRC achieved all of its safety goal strategic outcomes.

Performance Measures

Table 1 lists the agency's annual performance measures and their outcomes for the past 6 years. The performance measures quantify the agency's success in achieving its safety goal.

Table 1

FISCAL YEAR 2009 SAFETY GOAL PERFORMANCE MEASURES

Measure	2004	2005	2006	2007	2008	2009
1. Number of new conditions evaluated as red by the Reactor Oversight Process is ≤3.	1	0	0	0	0	0
2. Number of significant accident sequence precursors of a nuclear reactor accident is 0.	0	0	0	0	0	0
3. Number of operating reactors with integrated performance that entered the Manual Chapter 0350 process, or the multiple/repetitive degraded cornerstone column, or the unacceptable performance column of the Reactor Oversight Program action matrix, with no performance exceeding Abnormal Occurrence Criterion I.D.4 is ≤4.	1	0	0	1	0	0
4. Number of significant adverse trends in industry safety performance with no trend exceeding the Abnormal Occurrence Criterion I.D.4 is ≤1.	0	0	0	0	0	0
5. Number of events with radiation exposures to the public and occupational workers that exceed Abnormal Occurrence Criterion I.A is						
Reactors: 0	0	0	0	0	0	0
Materials: ≤3	0	1	0	0	0	0
Waste: 0	0	0	0	0	0	0
6. Number of radiological releases to the environment that exceed applicable regulatory limits is						
Reactor: ≤3	0	0	0	0	0	0
Materials: ≤2	1	0	0	0	0	0
Waste: 0	0	0	0	0	0	0

Analysis of FY 2009 Results

1. **Reactor Oversight Process:** The NRC reactor inspection program monitors nuclear power plant performance in three broad areas—reactor safety, radiation safety, and security and protection of the environment. Plant performance is analyzed based on many performance indicators and inspection findings. Each finding is then categorized into one of four categories—green, white, yellow, or red. Red findings indicate a finding of high safety significance. There were no red performance indicators or findings in FY 2009.

2. **Reactor significant precursors:** The second measure tracks significant precursor events. This statistical measure of risk determines the likelihood of an event impacting safety adversely. A significant precursor is an event that has a probability of 1 in 1,000 (or greater) of leading to substantial damage to the reactor fuel. No significant precursor events have been identified based on screening reviews.

3. **Reactor performance:** The conditions in this measure indicate whether the NRC finds significant performance issues in a plant during an inspection or from performance indicators under the Reactor

Oversight Program. If any of the conditions in this measure are met, the NRC will take action to ensure that plant safety is improved. There were no reactors that met the conditions in this measure.

4. Reactor safety trends: This measure tracks trends for several key indicators of industry safety performance. These indicators provide insights into major areas of reactor performance, including reactor safety, radiation safety, and emergency preparedness. The NRC applies statistical analysis techniques to each indicator to calculate long-term trends. These trends represent industry averages rather than individual plant performance. No statistically significant adverse trends have been identified in any of the indicators in FY 2009.

5. Nuclear material radiation exposures: This measure tracks the number of radiation exposures to the public and occupational workers that exceed Abnormal Occurrence Criterion I.A.3, which is defined as those events that produce unintended permanent functional damage to an organ or a physiological system, as determined by a physician. This measure tracks both nuclear reactors and other nuclear material users, such as hospitals and industrial users. No radiation exposures exceeding Abnormal Occurrence Criterion I.A.3 occurred in FY 2009.

6. Nuclear material releases to the environment: This measure indicates the effectiveness of the NRC's nuclear material environmental regulatory programs. Exceeding the applicable regulatory limits is defined as a total effective radiation dose equivalent to individual members of the public that is attributable to a licensed user of nuclear materials but does not exceed 0.1 rem in a year, exclusive of dose contributions from background radiation. No nuclear material releases to the environment that exceeded regulatory limits occurred in FY 2009.

The Industry Trends Program

The NRC measures the effectiveness of its nuclear reactor safety program activities based on the continued safe operation of the Nation's nuclear power plants. The NRC compiles data on overall safety performance using several industry-level performance indicators, a number of which are addressed in the following pages. These indicators show significant improvement in the long-term safety performance of nuclear power plants since 1993. Plant operating experience data have yielded a steady stream of improvements in the reliability of plant systems and components, plant operating procedures, training of power plant operators, and regulatory oversight. For ease of viewing, all the charts in this section display data since 1993.

The industry safety indicators are derived through engineering and scientific analyses by the NRC's Office of Nuclear Reactor Regulation and Office of Nuclear Regulatory Research. Since the final data are not available until February of each year, this report will only show final fiscal year data from FY 1993–2008. The results of these analyses are provided to the NRC Commission (SECY-09-0048) and reported annually to Congress.

Figure 10
SIGNIFICANT EVENTS

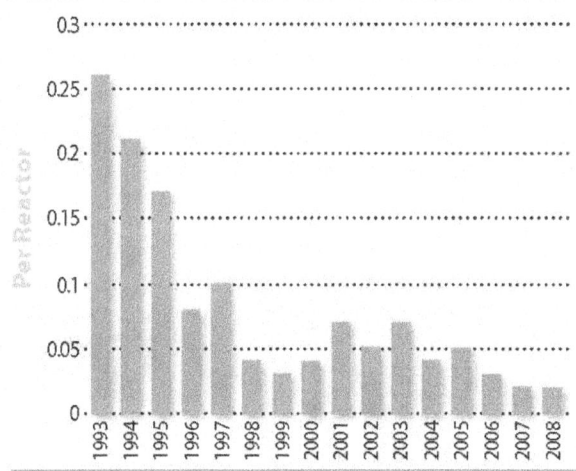

Significant events meet specific criteria such as degradation of important safety equipment. The agency reviews operating events and assesses their safety significance. The number of significant events has declined since 1993.

Figure 11
RADIATION EXPOSURE

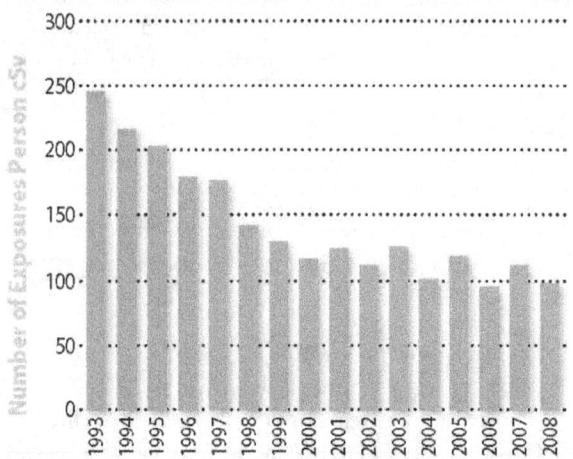

The total (collective) radiation dose received by workers is an indication of the radiological challenges of maintaining and operating nuclear power plants. The trend shows a reduction in collective dose since 1993 and demonstrates the effectiveness of the controls on radiation exposure implemented to meet these challenges.

Figure 13
SAFETY SYSTEM ACTUATIONS

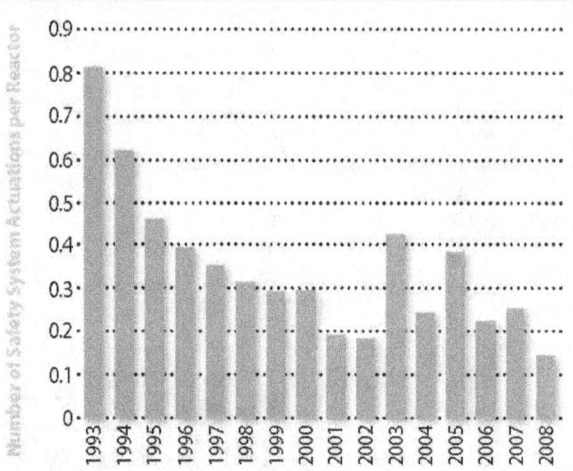

Safety systems mitigate off-normal events, such as the widespread power blackout in August 2003, by providing reactor core cooling and water addition. Actuations of safety systems that are monitored include certain emergency core cooling and emergency electrical power systems. Actuations can occur as a result of false alarms (such as testing errors) or in response to actual events.

Figure 12
AUTOMATIC SCRAMS

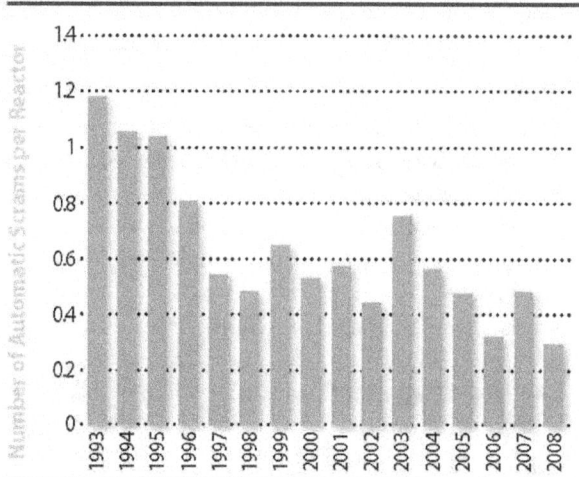

A scram is a basic reactor protection safety function that shuts down the reactor by inserting control rods into the reactor core. Scrams can result from events that range from relatively minor incidents to precursors of accidents. The massive power blackout in August 2003, accounts for most of the increase in FY 2003, but has not affected the statistical trend for number of scrams, which has been declining steadily since 1993.

Figure 14
PRECURSOR OCCURRENCE RATE

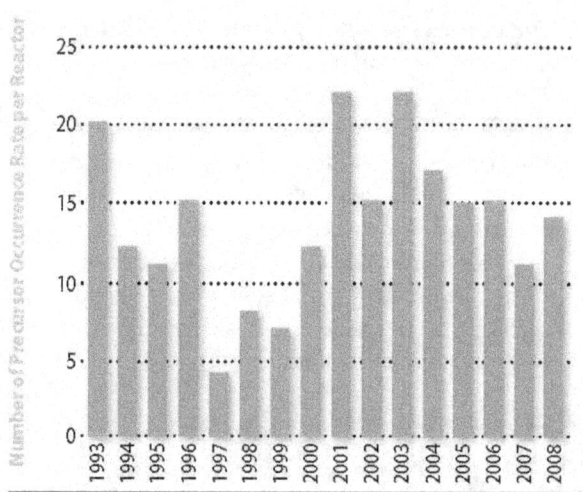

A precursor event is an event that has a probability of greater than 1 in 1 million of leading to substantial damage to the reactor fuel. The observed increase in precursors starting in FY 2001 is due to the increase

in Accident Sequence Precursor (ASP) Program scope (e.g., inclusion of external events and significance determination process findings) beginning in FY 2000. These increases in scope have resulted in the identification of an increasing number of lower-risk precursors (i.e., CCDP or ∆CDP <10⁻⁴). In addition, an increased number of outlier events (e.g., the 8 events leading to loss of offsite power due to the 2003 Northeast Blackout and the 11 events involving control-rod drive mechanism housing cracks between FY 2001 and FY 2003) account for the observed change. During the FY 2001 through FY 2008 period, the overall occurrence rate is statistically decreasing during the 8 year period. Due to the complexities associated with evaluating precursor events, the data always lag behind other indicators.

Figure 15
SAFETY SYSTEMS FAILURES

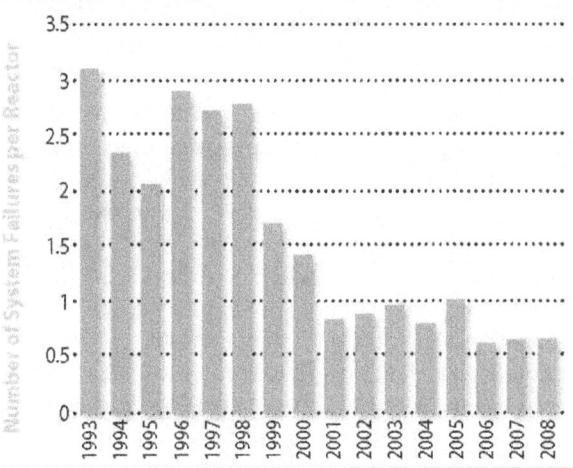

Safety system failures include any events or conditions that could prevent a safety system from fulfilling its safety function. The number of safety system failures across the industry has declined since 1993.

Nuclear Reactor Licensing Activity

The agency's nuclear reactor licensing activity ensures that licensees operate civilian nuclear power reactors and test and research reactors in a manner that adequately protects public health and safety and the environment while safeguarding special nuclear materials used in reactors. Safety at nuclear power plants has improved substantially over the past 20 years, as both the nuclear industry and the NRC have been proactive in identifying and correcting problems to improve the operation and maintenance of nuclear power facilities. The combined efforts of the nuclear industry and the NRC led to this improvement in the safety performance of nuclear power plants. For more information on reactor licensing, see http://www.nrc.gov/reactors/operator-licensing.html.

Licensing Actions

The NRC completed 1,002 reactor licensing actions in FY 2009. The agency has experienced a significant decrease in the number of licensing actions completed in the past 4 years (see Figure 16). This is predominately due to the fact that the agency received only 900 such actions in FY 2009, compared with an average of 1,250 submittals per year since 2003.

Figure 16
LICENSING ACTIONS

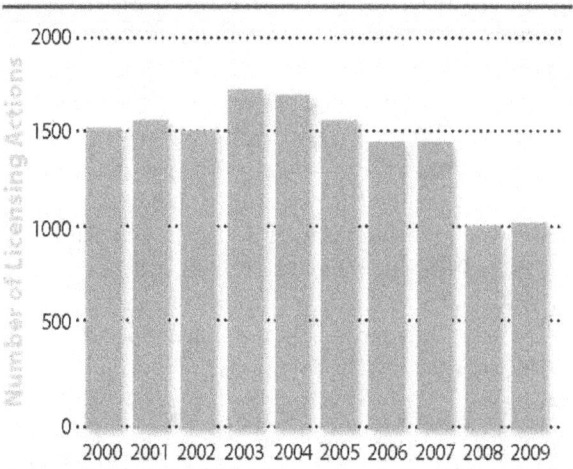

The NRC continues to complete licensing actions in a timely manner. The staff completed 94 percent of the licensing actions in the agency's inventory within 1 year of receipt and 100 percent within 2 years (see Figure 17).

Figure 17
LICENSING ACTION AGE

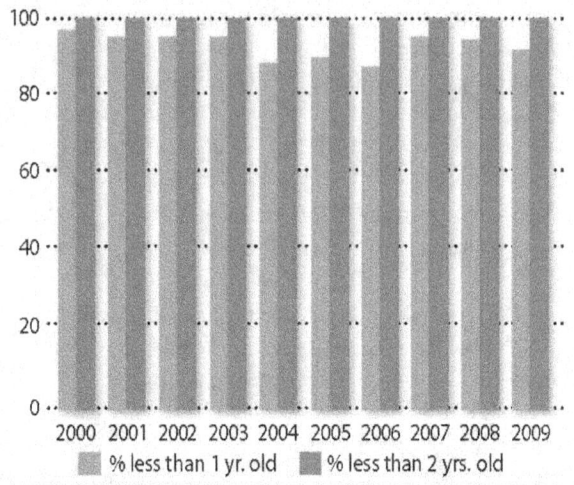

| 2000 2001 2002 2003 2004 2005 2006 2007 2008 2009 |
| ■ % less than 1 yr. old   ■ % less than 2 yrs. old |

Power Uprates

The NRC also evaluates nuclear reactor power uprate applications, which allow licensees to increase the power output of their plants. The NRC review focuses on the potential impacts of the proposed power uprate on overall plant safety and evaluates whether plant operation at the increased power level is safe. The cumulative additional power from all power uprates approved since 1977 is about 5,640 megawatts electric. The NRC currently has 11 power uprates under review. If approved, these uprates will add approximately 973 megawatts electric to the grid. The NRC expects to receive 40 new power uprate applications in the next 5 years. If approved, these uprates will add about 2,076 megawatts electric to the grid.

During FY 2009, the NRC undertook several rulemaking activities to improve protection of public health and safety and the environment and improve the regulatory framework. One rulemaking was the publication of a proposed rule to enhance NRC regulations for emergency preparedness. The agency also published a final rule on alternate fracture toughness requirements for protection against pressurized thermal shock events in reactor vessels using updated analysis methods, and a proposed rule that updates NRC requirements for the generic environmental impact statement (GEIS) addressing the environmental effects of renewing power reactor operating licenses.

New Reactor Licensing

The NRC revised Title 10 of the *Code of Federal Regulations* (10 CFR) Part 52, "Licenses, Certifications, and Approvals for Nuclear Power Plants," to require applicants for new nuclear power reactors to perform a design-specific assessment of the effects of the impact of a large, commercial aircraft. The rule requires applicants to use realistic analyses to identify and incorporate design features and functional capabilities to ensure, with reduced use of operator actions, that either the reactor core remains cooled or the containment remains intact and either spent fuel cooling or spent fuel pool integrity is maintained. In addition, the NRC issued interim staff guidance documents titled "Necessary Content of Plant Specific Technical Specifications" and "Generic Communication Plan for the Review of Combined License Applications." The communication plan outlines the internal and external communications required for various phases of the review, up to and including issuance of a license. Its purpose is to deliver clear and concise messages about new reactor requirements, as well as to convey the NRC's expectations and objectives to key internal and external stakeholders and other interested parties in a timely and efficient manner. For more information on new reactors, see http://www.nrc.gov/reactors/new-reactors.html.

New Reactor Designs

The NRC is actively reviewing several nuclear reactor designs, and plans to conclude these reviews with a design certification rulemaking. When an application references a certified design, the license application review can proceed that promotes safety and minimizes delays.

The NRC is currently performing the design certification review of the General Electric Economic Simplified Boiling-Water Reactor (ESBWR) design, the AREVA Evolutionary Power Reactor (EPR), and the Mitsubishi U.S. Advanced Pressurized-Water Reactor (USAPWR). The agency is also in the process of performing design certification amendment reviews for the Westinghouse AP1000 design and the General Electric Advanced Boiling Water Reactor (ABWR). In addition, vendors for four small reactors have requested preapplication discussions with the NRC. In FY 2009, the NRC has held public preapplication meetings with these vendors to help the NRC staff understand the designs of the various reactors.

In FY 2009, the staff issued an information paper describing plans to streamline the design certification rulemaking process. The staff evaluated this process as part of the NRC's Lean Six Sigma program in order to identify possible ways to shorten the rulemaking process and coordinate activities (design reviews, rulemaking, and licensing) to minimize the effect of the rulemaking on combined license (COL) schedules. As a result, the agency will improve the rulemaking process by adopting several internal enhancements that are expected to decrease the rulemaking review time by up to several months.

Early Site Permits

By issuing an early site permit, the NRC approves the site for a nuclear facility. Early site permits are valid for 10 to 20 years and can be renewed for an additional 10 to 20 years. The NRC review of an early site permit application addresses site safety issues, environmental protection issues, and plans for coping with emergencies, independent of the review of a specific nuclear plant design. The NRC issued early site permits to the Clinton site in Illinois on March 15, 2007; the Grand Gulf site in Mississippi on April 5, 2007; the North Anna site in Virginia on November 27, 2007; and the Vogtle site in Georgia on August 26, 2009.

Combined License

The goals for new reactors are to review COL applications, first to ensure the proposed new reactor design and planned operations will be in accordance with NRC regulations for safety, security, and the environment and second that the reviews will be completed on the schedules negotiated with applicants. For FY 2009, the NRC established a target to complete milestones associated with conducting 20 COL application reviews. To date, the NRC has docketed 18 COL applications from the nuclear power industry for sites across the country. Thirteen of the 18 applications are being actively reviewed. The NRC is developing the review schedule for the Turkey Point COL. In response to applicant requests, the NRC has suspended the reviews of the Grand Gulf, Victoria County, Callaway, Nine Mile Point, and River Bend COL applications.

In FY 2009, 9 COL applications were submitted to the NRC. The NRC developed a new set of goals to sequence project reviews, emphasizing those projects that are expected to complete licensing and construction and begin operation in the near term (potentially resulting in commercial operation in 2016–2017.

COMBINED LICENSE APPLICATIONS RECEIVED IN FY 2009

Site Name (Units)	State	Company	Accepted
Levy County (2 units)	FL	Progress Energy	10/06/2008
Victoria County (2 units)	TX	Exelon	10/30/2008
Fermi (1 unit)	MI	Detroit Edison	11/25/2008
Comanche Peak (2 units)	TX	Luminant Power	12/02/2008
River Bend (1 unit)	LA	Entergy	12/04/2008
Callaway (1 unit)	MO	AmerenUE	12/12/2008
Nine Mile Point (1 unit)	NY	UNISTAR	12/12/2008
Bell Bend (1 unit)	PA	PPL Generation	12/19/2008
Turkey Point (2 units)	FL	Florida Power & Light	9/02/2009

The NRC has developed a construction inspection program for plants to be licensed under 10 CFR Part 52, and undertook many critical development activities for this program in FY 2009. For example, the NRC produced a number of draft and final construction inspection program materials, such as inspection procedures, inspection strategy documents, regulatory guides, inspection manual chapters, and a construction inspection program information brochure for stakeholders in both English and Spanish. The staff developed an approach for maintaining completed inspections, tests, analyses, and acceptance criteria (ITAAC), and continued developing a detailed ITAAC closure verification process. NRC staff (1) continued inspector development and training, (2) deployed the initial version of the Construction Inspection Program Information Management System, which will capture inspection results and track ITAAC closure, (3) developed business processes to support additional identified information technology system needs, (4) continued development of generic inspection schedules, (5) continued development of enhancements to the existing assessment and enforcement program for new reactors, and (6) maintained an aggressive schedule of public meetings to provide a forum for stakeholders to participate and comment on staff proposals for ITAAC closure, licensee assessment, enforcement, and other construction inspection program topics.

The agency has in place a regular schedule of vendor inspections and an active program of international cooperation to support increased fabrication activities domestically and internationally in response to new reactor construction plans. The NRC conducts these inspections to ensure the effective implementation of quality assurance program requirements imposed on vendors by NRC applicants and licensees. The NRC conducts a minimum of 10 domestic and international routine and reactive vendor inspections per year. In FY 2009, 10 inspections were completed. The NRC held a highly successful "Workshop on Vendor Oversight for New Reactor Construction" with more than 600 participants from all program stakeholder groups. Related international cooperative efforts have included multinational vendor inspections, technical discussions with foreign regulatory counterparts, vendor experience and information sharing with other countries, NRC inspector rotations to facilities under construction in other countries, and participation in the Vendor Inspection Cooperation Working Group under the auspices of the Multinational Design Evaluation Program. Exchanges such as these have provided key insights into each country's methods of oversight and have enabled the NRC to build a foundation of trust and a rapport for communicating and sharing key information, findings, and enhancements to its own programs.

Advanced Reactor Program

The NRC has continued its efforts to support programs sponsored by the U.S. Department of Energy (DOE) such as the Generation IV Nuclear Energy Systems initiative, which is focused on research and development for a very-high-temperature reactor. Specifically, the NRC has concentrated on identifying and resolving generic policy issues as well as key technical issues for the licensing of a variety of advanced reactor designs. In addition, the agency continues to conduct preapplication interactions with private companies that are proposing small and medium-sized reactors for electrical and process heat applications.

License Renewal

The NRC grants nuclear reactor operating licenses for 40 years, which can be renewed for an additional 20 years. The review process for renewal applications is designed to assess whether a reactor can continue to operate safely during the extended period of operation.

To renew a license, the utility must demonstrate that the effects of aging will not adversely affect structures or components important to safety during the renewal period. Such structures and components include the reactor vessel, piping, electrical cabling, containment structure, and steam generators. For some structures or components, additional action may be needed to ensure adequate margins of safety. Additionally, the agency assesses the potential impacts of the extended period of operation on the environment.

The NRC has received applications to renew the licenses for 72 units at 43 sites since the license renewal program began in 2000 and has renewed licenses for 54 units at 31 sites during that time (see Figure 18). The NRC is currently reviewing applications to renew the licenses for 18 units at 12 sites. The agency expects that almost all of the licensees for currently licensed units will ultimately apply to renew their licenses.

In FY 2009, NRC achieved its goal of completing four application reviews, a major milestone under the license renewal program.

Figure 18
LICENSE RENEWAL APPLICATIONS

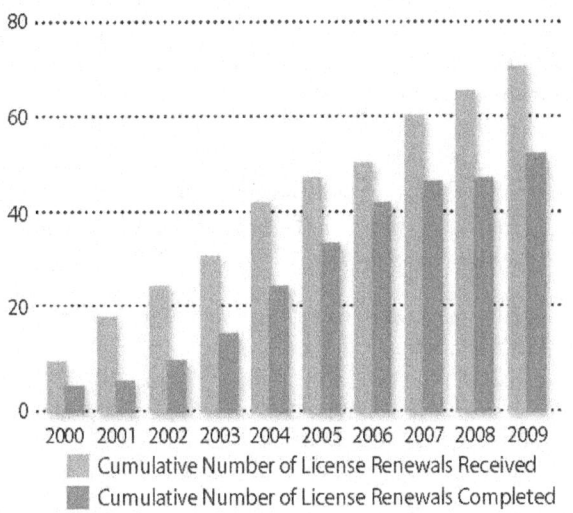

Cumulative Number of License Renewals Received
Cumulative Number of License Renewals Completed

Nuclear Reactor Inspection

The NRC's Reactor Oversight Process (ROP) outlines the agency's actions to verify that nuclear plants are being operated safely and in accordance with the NRC's rules and regulations. The NRC has full authority to demand that a licensee take immediate action for any conditions that result in excess risk to the public, including requiring a plant to shut down if necessary. The agency evaluates inspection findings and performance indicators to assess the safety performance of each operating nuclear power plant. The NRC performs a rigorous program of inspections at each plant and may perform supplemental

inspections and take additional actions to ensure that the plants address significant safety issues. The results of NRC inspection findings for each plant are available to the public at http://www.nrc.gov/nrr/oversight/assess/pim_summary.html. The NRC also conducts public meetings with licensees to discuss the results of the NRC's assessments of their safety performance.

In FY 2009, all of the Nation's nuclear power plants were operated in accordance with NRC safety and security requirements. In FY 2009, the safety indicators for nuclear plants as a whole showed no adverse trends and more than 99 percent of plant safety indicators were rated green, which is the highest safety rating.

The NRC continued to improve the ROP in FY 2009. Agency assessments confirm that the ROP has resulted in a more objective, risk-informed, and predictable regulatory process that focuses NRC and licensee resources on aspects of plant performance that have the greatest impact on safe plant operations. For more information on reactor inspection, see http://www.nrc.gov/reactors/operating.html.

Reactor Investigations and Enforcement

Compliance with NRC requirements plays an important role in giving the agency confidence that the licensee is maintaining safety. NRC policies deter noncompliance and encourage prompt identification and timely, comprehensive of safety corrective actions. Licensees, contractors, and their employees who do not achieve the high standard of compliance expected by the NRC are subject to enforcement sanctions. Each enforcement action depends on the circumstances of the case. The NRC will not permit licensees to continue to conduct licensed activities if they cannot achieve and maintain adequate levels of safety. In FY 2009, the agency took 30 escalated enforcement actions related to nuclear reactors and assessed $65,000 in fines. Allegations of reactor-related wrongdoing are referred to the Office of Investigations for appropriate action.

Fuel Facilities

The NRC licenses and inspects all commercial nuclear fuel facilities that process and fabricate uranium ore into reactor fuel. This fuel is the manufactured material that powers the Nation's nuclear reactors.

Licensing and inspection activities include detailed health, safety, safeguards, and environmental licensing reviews, as well as inspections of licensee programs, procedures, operations, and facilities to ensure safe and secure operations.

The NRC conducted several significant fuel cycle licensing reviews in FY 2009. The agency completed transfers of ownership to Babcock & Wilcox Nuclear Operations Group from BWX Technologies and from Nuclear Fuel Services. During FY 2009, the NRC issued 40-year license renewals to the AREVA Richland and Global Nuclear Fuels Americas fuel fabrication facilities. These were the first 40-year renewals approved by the agency under a policy established by the Commission in 2006. The basis for the extended renewals relies on the licensee's integrated safety analysis. The integrated safety analysis describes the management measures to ensure that the selected controls are available and reliable. The analysis allows a licensee to use risk information to identify hazards and to develop the engineered and human performance barriers relied on to control and mitigate hazards.

The NRC also completed its Report to Congress and issued renewed Certificates of Compliance for the United States Enrichment Corporation (USEC) gaseous diffusion plants located near Paducah, KY, and Piketon, OH. Gaseous diffusion is a technology used to produce enriched uranium by forcing gaseous uranium hexafluoride through special membranes. By using a large cascade of many stages, high separations can be achieved. Gaseous diffusion was the first economical enrichment process developed successfully. The agency held public meetings near both of the facilities to allow for public input on the certificate renewal process. The NRC previously renewed the gaseous diffusion plant certificates in 2003.

To support growing industry interest in the potential recycling and reprocessing of spent nuclear fuel, the NRC staff continued its efforts to revise the regulatory framework for reprocessing. These efforts included completing a regulatory gap analysis, conducting a public meeting, and considering stakeholder feedback. In the regulatory gap analysis, the staff identified approximately 20 areas or "gaps" in its current regulations that the NRC must or should address in order to establish an effective and efficient regulatory framework for licensing a reprocessing facility.

Working with the U.S. Department of Commerce, U.S. Department of State, and DOE, the NRC supported the implementation of the additional protocol to the U.S./International Atomic Energy Agency (IAEA) Safeguards Agreement (Additional Protocol) by revising 10 CFR Part 75, "Safeguards on Nuclear Material—Implementation of US/IAEA Agreement," and collecting and reviewing data from licensees subject to the additional protocol.

The NRC received two applications for COL licenses for uranium enrichment facilities. The first, submitted in December 2008 by AREVA, is for a centrifuge enrichment facility to be built near Idaho Falls, ID. The second, submitted in June 2009 by General Electric-Hitachi, is for a laser-based enrichment facility to be built in Wilmington, NC. The NRC has completed the initial environmental review scoping effort, including conducting several public meetings in the vicinity of the proposed facilities and meeting with affected State, local, and tribal officials.

The agency also sponsored the fourth annual Fuel Cycle Information Exchange (FCIX) conference. The FCIX involved multiple presentations by staff, industry, and stakeholders on various regulatory aspects of the nuclear fuel cycle, including a reprocessing recycling workshop with a multinational panel of experts. This was the most successful FCIX to date, with more than 250 participants. For more information on fuel facilities, see http://www.nrc.gov/materials/fuel-cycle-fac.html.

Investigation and Enforcement

Compliance with NRC requirements plays an important role in giving the agency confidence that safety is being maintained. NRC policies deter noncompliance and encourage prompt identification and timely, comprehensive corrective actions. Licensees, contractors, and their employees who do not achieve the high standard of compliance expected by the NRC are subject to enforcement sanctions. Each enforcement action depends on the circumstances of the case. The NRC will not permit licensees to continue to conduct licensed activities if they cannot achieve and maintain adequate levels of safety. In FY 2009, the NRC took 10 escalated enforcement actions related to fuel facilities with $32,500 in fines assessed.

Nuclear Material Users

The NRC licenses and inspects the commercial use of nuclear material for industrial, medical, and academic purposes. Commercial uses of nuclear materials include medical diagnosis and therapy, medical and biological research, academic training and research, industrial gauging and nondestructive testing, production of radiopharmaceuticals, and fabrication of commercial products (such as smoke detectors) and other radioactive sealed sources and devices. The NRC and 37 Agreement States regulate more than 22,500 specific materials licensees and 150,000 general materials licensees. The NRC currently regulates and inspects approximately 2,970 specific licensees for the use of nuclear byproduct and other radioactive materials. The NRC also expects to complete 2,900 materials licensing actions and 1,200 routine health and safety inspections.

In FY 2009, the NRC deployed the National Source Tracking System, a centralized national registry that provides lifetime accounting of certain high-risk radioactive materials used in industry, medicine, and research. Licensees had to begin using the system by January 31, 2009.

Virginia became an Agreement State on March 31, 2009, and New Jersey became the 37th Agreement State on September 30, 2009. Agreement States assume regulatory responsibility over certain types and small quantities of nuclear material. These two new Agreement States will take over regulatory responsibility for approximately 900 materials licensees.

Detailed health and safety reviews of license applications, as well as inspections of licensee procedures, operations, and facilities, provide reasonable assurance of safe operations and the production of safe products. The NRC routinely inspects nuclear materials licensees to ensure that they are using nuclear materials safely, maintaining accountability of those materials, and protecting public health and safety. The agency also analyzes operational experience from NRC and Agreement State licensees and regularly evaluates the safety significance of events reported by licensees and Agreement States.

In FY 2009, the NRC completed reviews of 2,726 materials licensing actions and 1,091 materials program inspections. From 2003 through 2009, the NRC has maintained the timeliness of its reviews of nuclear materials license renewals and sealed source and device designs. In addition, the NRC completed 97 percent of new application and license amendment reviews within 90 days and 91 percent of the requests for license renewal and sealed source and device design reviews within 180 days of receipt.

Figure 19

TIMELINESS REVIEW OF NUCLEAR MATERIAL LICENSING APPLICATIONS

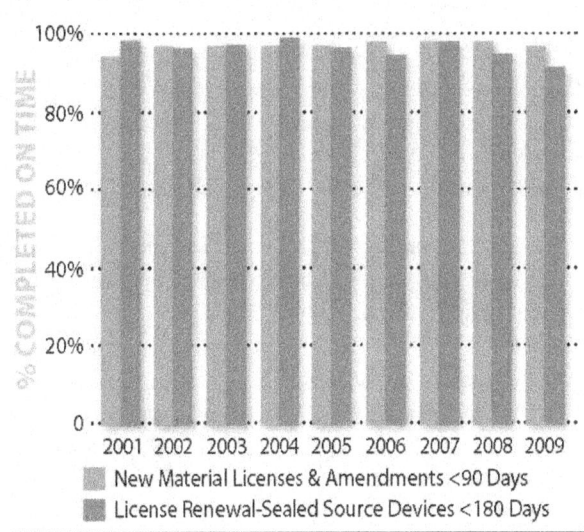

The NRC also works with international counterparts, both bilaterally and through multilateral organizations, to enhance the safety and security of radioactive sources. Examples of these activities include participating in ongoing meetings of countries implementing the IAEA Code of Conduct on the Safety and Security of Radioactive Sources to ensure harmonized national approaches, and working bilaterally with countries of the Commonwealth of

Independent States to support regulatory control over high-risk sources of concern. (See the section on international activities for additional details.) For more information on nuclear material users, see http://www.nrc.gov/materials/ql-materials.html.

Rulemaking Activities

In FY 2009, the NRC undertook several rulemaking activities to allow the use of radioactive materials while protecting public health and safety and the environment. These activities included rulemaking to enhance domestic nonproliferation activities in accordance with IAEA recommendations, implementing improvements to the licensing and distribution of byproduct materials, revising the requirements for categorical exclusion from environmental review, and amending the preceptor attestation requirements for medical licensees. The agency also published several rules related to certificates of compliance that certify the safety of casks for the storage of spent nuclear fuel, under 10 CFR Part 72, "Licensing Requirements for the Independent Storage of Spent Nuclear Fuel, and High-Level Radioactive Waste, and Reactor-Related Greater Than Class C Waste," including NAC-UMS Amendment 5; MAGNASTOR; HI-STORM100 Amendment 6; and NUHOMS Amendment 10.

The NRC is updating 10 CFR Part 110, "Export and Import of Nuclear Equipment and Material" to revise the definition of radioactive waste, incorporate changes to 10 CFR Part 110, Appendix P, "Category 1 and 2 Radioactive Material" based on experience gained throughout 2005-2008, and rewriting/clarifying 10 CFR 110.23, "General License for the Export of Byproduct Material," for the export of byproduct material. The proposed rule was approved by the Commission and published for comment in the *Federal Register* in June 2009.

High-Level Waste Repository

In FY 2008, DOE submitted a license application to the NRC seeking authorization to construct a geologic repository at Yucca Mountain, NV. The NRC formally docketed the DOE license application for the proposed high-level waste repository and determined that it was

practicable to adopt DOE's final environmental impact statement, subject to further supplementation.

In FY 2009, NRC technical staff began conducting a safety review of the license application. The agency is holding an impartial hearing proceeding as part of the licensing process. On October 17, 2008, the Commission issued a "Notice of Hearing and Opportunity To Petition for Leave To Intervene on an Application for Authority To Construct a Geologic Repository at a Geologic Operations Area at Yucca Mountain." It published this notice in the *Federal Register* on October 22, 2008 (73 FR 63029). In December 2008, the State of Nevada; the State of California; the Nuclear Energy Institute; Inyo County, California; the Timbisha Shoshone Tribe; the Native Community Action Council; the Timbisha Shoshone Yucca Mountain Oversight Program Non-Profit Corporation; and the Caliente Hot Springs Resort filed petitions for leave to intervene, requests for hearings, and contentions. In addition, Inyo County in California and the Nevada counties of Nye, Churchill, Esmeralda, Lander, and Mineral (jointly); Clark; and White Pine also filed petitions for leave to intervene. On December 22, 2008, Eureka and Lincoln counties in Nevada filed requests to participate as interested governmental participants. The Commission received 322 contentions.

The NRC's Chief Administrative Judge established three licensing boards, called Construction Authorization Boards, to preside over the proceeding. From March 12, 2009, through April 2, 2009, the construction authorization boards conducted prehearing conferences with parties, petitioners for intervention, and interested governmental participants. On May 11, 2009, the three Atomic Safety and Licensing Board Panel Construction Authorization Boards issued an order admitting eight parties and 299 contentions.

The NRC published final regulations (74 FR 10811) on March 13, 2009, implementing the U.S. Environmental Protection Agency's (EPA's) revised standards for doses that could occur after 10,000 years, but within the period of geologic stability. The final rule also specifies a range of values for the deep percolation rate to be used to represent climate change after 10,000 years, as called for by EPA, and specifies that calculations of radiation doses for workers use the same weighting factors that EPA is using for calculating individual doses to members of the public. The final rule became effective on April 13, 2009.

The NRC continued to interact with DOE to assess technical and regulatory issues related to its spent fuel management program, which will use standardized transportation, aging, and disposal (TAD) canisters. The NRC received two TAD applications in late FY 2009, requesting approval for storage and transportation. The TAD canister will be the primary means for packaging spent nuclear fuel for interim storage and for transportation to and disposal in the proposed repository at Yucca Mountain. For more information on high level waste, see http://www.nrc.gov/waste/hlw-disposal.html.

Spent Fuel Storage and Transportation

The NRC ensures that spent fuel is safely stored to support continued reactor operations and is safely transported when necessary. The NRC conducts licensing and certification reviews to ensure (1) storage facility and cask design compliance with NRC regulations for storage of spent fuel and (2) safe transport of domestic and international shipments of spent fuel and other risk-significant radioactive materials.

Licensees safely and securely transport shipments of radioactive materials each year within the United States. Several Federal agencies share responsibility for regulating the safety and security of those shipments. The NRC closely coordinates its transportation-related activities with those of the U.S. Department of Transportation and, as appropriate, DOE. To help ensure the safety and security of both spent fuel storage and transportation, the NRC inspects vendors, fabricators, and licensees using transport packages, spent fuel storage casks, and interim storage of spent fuel both at and away from reactor sites.

In FY 2009, the NRC completed 93 transport package design reviews and 17 storage cask and installation design reviews. The NRC review of transportation and interim storage licensing requests ensures that shipments are made in NRC-approved packages that

meet rigorous performance requirements and verifies that spent fuel is safely stored, thereby enabling continued reactor and decommissioning operations. The NRC also conducted 17 inspections of activities related to radioactive material package certificate holders, spent fuel storage cask certificate holders, and conducted dry run inspections at independent spent fuel storage facilities to ensure that licensees design, fabricate, and use casks according to approved safety requirements.

Figure 20
STORAGE AND TRANSPORTATION DESIGN REVIEWS COMPLETED

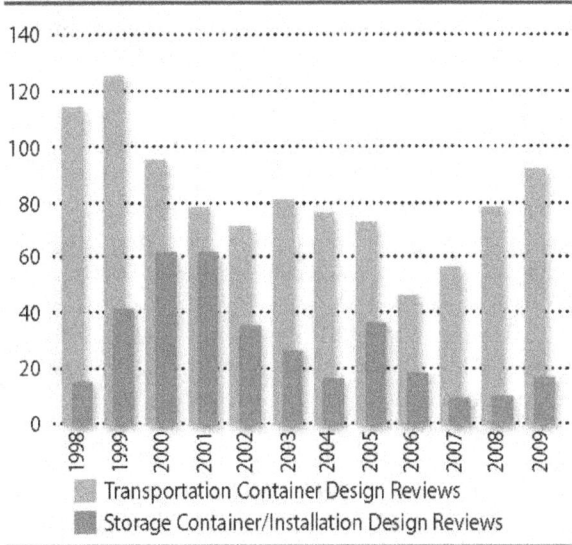

Transportation Container Design Reviews

Storage Container/Installation Design Reviews

The NRC developed an effective strategy to address the pending expiration of numerous transportation certificates of compliance on October 1, 2008. This strategy helped to ensure public health and safety by creating a mechanism for continued shipment of certain materials using the existing radioactive material transportation packages. The NRC received and approved 15 applications from vendors and shippers requesting limited continued use of expired packages.

The NRC held two public workshops with the Nuclear Energy Institute (NEI) and the regulated industry to improve communications and clarify expectations on technical topics. The first public workshop focused on shielding and radiation protection requirements and the methods used to demonstrate compliance with transportation and spent fuel storage regulations. The agency held a second public workshop on licensing and certification process improvements. The results of the workshop included a discussion of the draft acceptance review procedure. The NRC issued the draft procedure for public review and comment while the staff began piloting its use. Many vendors and licensees attended the workshops and provided very effective and positive feedback.

Investigation and Enforcement

Compliance with NRC requirements plays an important role in giving the agency confidence that safety is being maintained. NRC policies deter noncompliance and encourage prompt identification of issues and timely, comprehensive corrective actions. Licensees, contractors, and their employees who do not achieve the high standard of compliance expected by the NRC are subject to enforcement sanctions. Each enforcement action depends on the circumstances of the case. The NRC will not permit licensees to continue to conduct licensed activities if they cannot achieve and maintain adequate levels of safety. In FY 2009, the NRC took 62 escalated enforcement actions related to nuclear materials users and assessed $69,000 in fines. Allegations of materials-related wrongdoing are referred to the Office of Investigations for appropriate action.

Decommissioning and Low-Level Waste

In FY 2009, the NRC provided oversight of decommissioning activities at approximately 70 power and early demonstration reactors, research and test reactors, uranium recovery sites, and complex materials sites and fuel cycle facilities. Decommissioning removes radioactive contamination from buildings, equipment, ground water, and soil, achieving levels that permit the release of the property, with or without restrictions on its future use by the public. The NRC terminates the licenses for decommissioned facilities after the licensees demonstrate that the residual onsite radioactivity is within regulatory limits and sufficiently low to protect public health and safety and the environment. Completion of decommissioning, environmental, and performance assessment activities enables sites to

return to productive use while ensuring that residual radioactivity does not pose an unacceptable risk to the public.

In addition to the uranium recovery sites undergoing decommissioning, the NRC conducts regulatory oversight at five operational uranium recovery sites and reviews and approves the applications for new, restarting, or expanding uranium recovery facilities. Additionally, the NRC conducted a number of regulatory activities to help ensure the safe management and disposal of the low-level radioactive waste generated by radioactive material users, nuclear power plants, and other NRC licensees. The agency performed monitoring visits and issued reports for the Savannah River Site Saltstone facility and Idaho National Laboratory.

In late September 2008, the NRC approved the application for restarting the COGEMA/Christiansen Ranch uranium recovery facility. In FY 2009, the NRC completed decommissioning activities at the Rancho Seco power reactor and the Sigma-Aldrich materials facility. Also in FY 2009, the NRC continued to review seven applications for new, expanding, or restarting uranium recovery facilities received in FY 2007, including initiating environmental reviews. The agency issued the final generic environmental impact statement for in situ uranium recovery facilities in June 2009. In FY 2009, the NRC completed the second annual "Waste Incidental to Reprocessing Monitoring Report." For more information on decommissioning, see http://www.nrc.gov/about-nrc/regulatory/decommissioning.html.

Research Activities

The NRC's safety research program evaluates and resolves safety issues for nuclear power plants and other facilities regulated by the NRC. The agency conducts its research program to evaluate existing and potential safety issues; supply independent expertise, information, and technical judgments to support timely and realistic regulatory decisions; reduce uncertainties in risk assessments; and develop technical regulations and standards. When possible, the NRC engages in cooperative research with other Government agencies, the nuclear industry, universities, and international partners.

During the past year, the NRC research program has addressed key areas that support the agency's safety mission. Some of the more important issues addressed include the verification and validation of fire safety models, material degradation of reactor system and pressure boundary components, new digital instrumentation and control systems, seismic hazard issues, and severe reactor accident consequence analyses.

Fire Safety

During FY 2009, the NRC's fire safety research program focused on risk-informed fire protection activities, which endorse National Fire Protection Association Standard 805. Work has also continued on fire modeling activities, including a fire modeling users' guide for nuclear power plant applications. NRC fire safety research continues to focus on fuel cycle issues and the performance of spent nuclear fuel transportation cask seals in beyond-design-basis fires.

Advanced Reactor Research

The NRC has initiated research activities in a number of major technical areas related to licensing a prototype high-temperature gas-cooled reactor that can be used to generate electricity and hydrogen.

Materials Degradation

The NRC continues to conduct research on materials degradation issues for currently licensed reactors. The purpose of this research is to identify susceptible materials and assess component-specific degradation mechanisms to ensure continued safe operation. The staff is also performing research on reactor internals to determine the effects of neutron fluence and thermal effects on the physical properties of reactor internal materials.

Digital Instrumentation and Controls

The NRC is actively engaged in research to support the licensing of new digital instrumentation and control systems intended for use in retrofits to operating reactors and in new and next-generation reactors. The NRC is also actively engaged in ongoing research associated with the evaluation of digital systems for cyber vulnerabilities.

Seismic Research (Earth Sciences)

The NRC is conducting research on seismic hazard issues to support the siting of new reactors and the evaluation of the seismic safety of existing nuclear facilities. In cooperation with academic institutions, other Federal and State agencies, and industry, the NRC is conducting a program to develop ground motion propagation and earthquake source zone models. In cooperation with the U.S. Geological Survey and the National Oceanic and Atmospheric Administration, the NRC is also conducting a study of potential tsunami sources and the potential hazards to NRC-regulated facilities.

State-of-the-Art Reactor Consequence Analysis

The NRC, the U.S. nuclear industry, and the international nuclear community have performed extensive severe accident research to improve their understanding of the phenomena of severe accidents; the performance of plant systems and components under these conditions; the timing, magnitude, and composition of the radioactive material release; the effectiveness of the different design and mitigative measures, including emergency preparedness; and the understanding of the effects of radiation exposure on humans.

Emergency Preparedness and Incident Response

NRC emergency preparedness and incident response activities ensure that adequate measures can and will be taken to mitigate plant events and to minimize possible radiation doses to members of the public, and ensure that the agency can respond effectively to events at its licensees' sites.

The agency is currently engaged in a rulemaking effort that proposes to update the emergency preparedness regulations. Enhancements to the regulations include codifying voluntary industry efforts since September 11, 2001. The NRC issued the proposed rule in the *Federal Register* on May 18, 2009, and extended the public comment period, based on stakeholder feedback, to October 19, 2009. In June 2009, the NRC held 11 joint public meetings with the Federal Emergency Management Agency

(FEMA) on the proposed rule and associated draft guidance documents in each NRC region and near NRC headquarters. During these public meetings, the NRC demonstrated its openness in the regulatory process and increased overall stakeholder involvement by hosting the meetings in both live and Web-based meeting formats using new technologies. Participants from all over the country who did not have the opportunity to travel to the meeting were able to participate effectively, using those new technologies.

In 2009 the NRC continued to work with States to address the replenishment of potassium iodide supplies, to be used as a supplement to public protective actions, within the 10-mile emergency planning zones around nuclear power plants. The Commission has decided to modify its potassium iodide distribution policy from a one-time replenishment to one providing the tablets to affected States that request them and then replenishing stockpiles upon States' requests, consistent with the tablet shelf life.

In FY 2009, the NRC maintained its headquarters Operations Center and modernized two legacy computer systems in order to improve functionality, enhance cyber security, and reduce operating costs. The new systems provide needed capabilities for handling multiple, ongoing events and provide the ability to quickly and accurately share information with NRC responders in the regions, at sites, or at Headquarters. The NRC also continued its modernization of the Emergency Response Data System, which provides real-time information from nuclear power plants during events to the NRC and State emergency operations centers. The modernization of this system includes improvements to the user interface, accessibility via the Internet, and enhanced cyber security.

In FY 2009, the NRC engaged in multiple emergency exercises with its licensees and Federal partners. NRC emergency responders participated in 13 exercises with licensee sites across the county, 4 of which involved the NRC headquarters response team. These exercises focused on the implementation of onsite and offsite radiological emergency plans by the licensee, as well as State and local responders. The NRC also

uses exercises to train its response organization and to practice coordination activities with Federal partners, including the U.S. Department of Homeland Security. The NRC participated in hostile-action-based (HAB) emergency preparedness drills at the Three Mile Island and Turkey Point stations and coordinated with FEMA to observe numerous other HAB drills conducted as part of industry's voluntary program to gain a better understanding of the unique challenges that hostile-action events pose.

In addition to exercises involving its licensees, the NRC participated in Federal emergency exercises during FY 2009, including the National-Level Exercise (NLE 09), the annual Continuity Exercise (Eagle Horizon 09), and a Federal Radiological Monitoring and Assessment Center field exercise (Empire 09).

The 2009 H1N1 influenza outbreak provided an opportunity to revisit the agency's plans and prioritize improvements for implementation in advance of the next flu season. The agency used Federal guidance to shape preplanned actions for the protection of the NRC workforce. The NRC worked with the nuclear industry to coordinate plans, with the goal of ensuring that the nuclear sector is prepared to address the challenges of a pandemic and maintain the standards of safety and security required for operations. For more information on emergency preparedness, see http://www.nrc.gov/about-nrc/emerg-preparedness. html.

International Activities

The NRC's international efforts include participation in activities that support U.S. Government compliance with international treaties and agreements; export/import licensing of nuclear facilities, equipment, and materials; programs of bilateral nuclear cooperation and assistance; and support for multinational nuclear safety organizations such as IAEA and the Organisation for Economic Co-operation and Development's Nuclear Energy Agency.

Notable accomplishments in FY 2009 in the area of international treaties and agreements include high-level NRC participation in the May 2009 Review Meeting of Contracting Parties to the Joint

Convention on the Safety of Spent Fuel Management and the Safety of Radioactive Waste Management, and Commission review of U.S. Government agreements for peaceful uses of nuclear energy with the United Arab Emirates. The Commission completed work with NRC licensees and rulemaking activities to finalize the initial reporting requirements for compliance with the U.S. Protocol Additional to the Agreement between the United States and the IAEA.

In the area of export/import licensing, the NRC continued to work both domestically and internationally to enhance nuclear safety and security through the regulatory oversight of radioactive sources (see the section on nuclear materials users for specific examples). In July 2009, the NRC attended an IAEA open-ended meeting of technical and legal experts to share information on lessons learned from implementation of the "Supplementary Guidance on Import and Export of Radioactive Sources."

Accomplishments in the area of bilateral activities during FY 2009 include the first steps in implementing the information exchange arrangement with the National Nuclear Safety Administration of China and a similar information exchange arrangement with the Vietnam Agency for Radiation and Nuclear Safety and Control.

The NRC continues to support the development and implementation of programs focused on leveraging the knowledge and resources within the international regulatory community in the licensing of new reactor designs. In the multilateral context, the NRC continues its leadership role in the Multinational Design Evaluation Program, through which regulatory authorities in over a dozen countries share expertise and resources in reviewing new and future reactor designs. Currently, the program consists of three issue-specific and two design-specific working groups. The Digital Instrumentation and Controls Working Group, led by the United States, established common positions in digital instrumentation and control system design. The Vendor Inspection Cooperation Working Group has conducted several parallel inspections that involved more than one regulator, and the Codes and

Standards Working Group is nearing completion of a project to compare the pressure boundary codes of four member countries. The design-specific working groups, based on the Westinghouse AP1000 and the AREVA EPR designs, also each established three sub-working groups. In FY 2009, the Policy Group, which is the governing body of the program, extended the commitment to the program from 2 to 5 years while requiring that each working group achieve significant interim results.

The NRC has worked both domestically and internationally (bilaterally and multilaterally) to enhance nuclear safety and security through the regulatory oversight of radioactive sources. In FY 2009, the NRC expanded assistance efforts related to radioactive sources for the regulatory authorities of the Commonwealth of Independent States, expanded assistance provided to the Iraqi Radioactive Source Regulatory Authority, established initial assistance efforts for select regulatory authorities in Africa, and enhanced support for and coordination with source-related assistance activities conducted by IAEA. The NRC has also worked with other U.S. agencies and IAEA to develop international security guidance documents for materials control, accounting, and physical protection.

The NRC participated in a working group with representatives of DOT and the Canadian Nuclear Safety Commission to publish in March 2009 NUREG-1886, "Joint Canada-United States Guide for Approval of Type B(U) and Fissile Material Transportation Packages." This NUREG will provide the framework for U.S. and Canadian cooperation and acceptance of each country's Type B(U) and fissile materials transportation package design approvals for export and import. The NRC expects to sign a Memorandum of Understanding by the end of FY 2009 for joint implementation of the Guide.

The NRC has put considerable effort into bilateral inspection training activities, especially with Finland, Japan, South Korea, and Taiwan, as these countries are building new reactors and/or are the site of major nuclear manufacturing facilities. Future cooperation with the regulatory bodies in these countries is expected as more vendors become active in the nuclear market.

STRATEGIC GOAL 2: SECURITY

Ensure Adequate Protection in the Secure Use and Management of Radioactive Materials

Strategic Outcome

The NRC has the following strategic outcome associated with the agency's goal to ensure the secure use and management of radioactive materials: Prevent any instances where licensed radioactive

materials are used domestically in a manner hostile to the security of the United States.

RESULTS: In FY 2009, the NRC achieved its security goal strategic outcome.

Performance Measures

The table below lists the agency's annual performance measures and their outcomes for the past 6 years. The performance measures are used to determine the agency's success in achieving its security goal. The NRC met all of the FY 2009 security goal performance measure targets.

FY 2009 Security Goal Performance Measures						
Measure	2004	2005	2006	2007	2008	2009
1. Number of unrecovered losses or thefts of risk-significant radioactive sources is zero.	0	0	0	0	0	0
2. Number of substantiated cases of theft or diversion of licensed, risk-significant radioactive sources or formula quantities of special nuclear material, or attacks that result in radiological sabotage is zero.	0	0	0	0	0	0
3. Number of substantiated losses of formula quantities of special nuclear material or substantiated inventory discrepancies of formula quantities of special nuclear material that are judged to be caused by theft or diversion or by substantial breakdown of the accountability system is zero.	0	0	0	0	0	0
4. Number of substantial breakdowns of physical security or material control (i.e., access control containment or accountability systems) that significantly weaken the protection against theft, diversion, or sabotage is less than or equal to one.	0	0	0	0	0	0
5. Number of significant unauthorized disclosures of classified and/or safeguards information is zero.	0	0	0	0	0	0

Analysis of FY 2009 Results

1. **Unrecovered losses or thefts:** This measure includes any loss or theft of radioactive nuclear sources that the NRC has determined to be risk significant. The measure tracks the NRC's performance in ensuring that those radioactive sources that the agency has determined to be risk significant for public health and safety are accounted for at all times. The ability to account for these sources is vital to securing the Nation's critical infrastructure from "dirty bomb" attacks or other means of radioactive material dispersal. There was no loss or theft of radioactive nuclear material that the NRC determined to be risk significant during FY 2009.

2. **Thefts or diversion:** This measure includes whether NRC-licensed facilities maintain adequate protective capabilities to prevent theft or diversion of nuclear material or sabotage that could result in harm to public health and safety. There were no substantiated cases of theft or diversion of licensed, risk-significant radioactive sources or formula quantities of special nuclear material or attacks that resulted in radiological sabotage during FY 2009.

3. **Loss or inventory discrepancy:** This measure includes whether special nuclear material is accounted for at all times and whether any losses of this material occur that could lead to the creation of an improvised nuclear device or other type of nuclear device. Furthermore, the measure tracks whether the systems in place at NRC-licensed facilities maintain accurate inventories of special nuclear material that the facilities process, use, or store. There were no substantiated losses of formula quantities of special nuclear material or substantiated inventory discrepancies of formula quantities of special nuclear material that were caused by theft or diversion or by substantial breakdown of the accountability system during FY 2009.

4. **Substantial breakdowns of physical security:** This measure includes any breakdowns in access control, containment, or accountability systems that significantly weakened the protection against theft, diversion, or sabotage for nuclear materials that the Commission has determined to be risk significant. There were no substantial breakdowns of physical security during FY 2009.

5. **Significant unauthorized disclosures:** This measure includes significant unauthorized disclosures of classified and/or safeguards information that cause damage to national security or public safety. This measure tracks whether information that can harm national security (classified information) or cause damage to public health and safety (safeguards information) has been stored and used in such a way as to prevent its disclosure to the public, terrorist organizations, other nations, or personnel without a need to know. There were no significant disclosures that caused damage to national security or public safety during FY 2009.

Nuclear Security Activities

The NRC must remain vigilant of the security of nuclear facilities and materials. The agency achieves its goal of protecting the common defense and security by using licensing and oversight programs similar to those employed in achieving its safety goal. The NRC aims to allow licensees to realize the benefits of nuclear materials through their secure use, while at the same time placing only those regulatory requirements that are necessary on those licensees. Listed below are noteworthy NRC accomplishments in the area of nuclear security. For more information on nuclear security, see http://www.nrc.gov/security/domestic.html.

Security Inspections

The NRC continued to maintain vigilant oversight of security in the nuclear industry and to implement the agency's security procedures. There were no substantial breakdowns of physical security at any commercial nuclear power plant, as determined by the NRC's implementation of its baseline security inspection program. This inspection effort resides within the "security cornerstone" of the agency's ROP. The security cornerstone focuses on the following five key licensee performance attributes: access authorization, access control, physical protection systems, material control and accounting, and response to contingency events. Through the results obtained from all oversight activities, including baseline security inspections and performance indicators, the NRC determines whether licensees comply with requirements and can provide high assurance of adequate protection against the design-basis threat for radiological sabotage.

The NRC regularly carries out force-on-force inspections at commercial operating nuclear power plants and Category I fuel cycle facilities as part of its comprehensive security program. The agency uses these inspections to evaluate the effectiveness of plant security programs to prevent radiological sabotage. The agency conducts force-on-force inspections at least once every 3 years at each commercial nuclear power plant and Category I fuel facility.

Force-on-force inspections assess a nuclear facility's ability to defend against the design-basis threat, which characterizes the adversary against which plant owners must design appropriate defenses, such as physical protection systems and response strategies. A force-on-force inspection includes both tabletop drills and simulated combat between a mock commando-type adversary force and the nuclear facility's security force. In FY 2009, the agency completed 11 force-on-force inspections and submitted the fourth annual Report to Congress on the results of the security inspection program.

The agency also pursued recommended enhancements to its allegation and inspection programs based on a lessons-learned review that followed an agency investigation into reports of inattentive security officers at the Peach Bottom nuclear power plant in Pennsylvania. To address lessons learned, on December 29, 2008, the agency issued revised guidance on contacting allegers, engaging licensees with requests for information, and independently validating licensee inputs, among other things.

The NRC worked with DOE to recover unwanted or orphaned radioactive sources. The source recovery program aids in preventing inadvertent source disposal during recycling, and guards against

malevolent use of sources. Since the inception of this program in 1997, more than 17,700 radioactive sources have been recovered from more than 690 sites within the United States. In FY 2009, over 2,000 radioactive sources were recovered.

The NRC is assisting U.S. Customs and Border Protection in fulfilling its congressional mandate to verify the legitimacy of radioactive material shipments coming into the United States through established ports of entry. The NRC regularly provides Customs and Border Protection information on the licensing of radioactive materials, including import and export licensing data, and has established processes to provide around-the-clock technical support for the verification of licensing status for materials in transit.

In addition to continuing to evaluate the need to enhance security at byproduct material licensees in FY 2009, the NRC inspected licensee compliance with the safety and security measures and coordinated with Agreement States to identify and resolve any implementation issues. The NRC also issued security orders to irradiator facilities, manufacturer and distributor facilities, and licensees shipping IAEA Category 1 quantities, including orders requiring this group of licensees to implement a program to fingerprint and conduct criminal history checks for access to safeguards information and material. The NRC and Agreement States issued orders and legally binding agreements to licensees subject to increased controls that require fingerprinting and criminal history checks for access to material. The NRC and Agreement States will continue to inspect these licensees to ensure the proper implementation of the increased control orders and other associated requirements. The NRC revised its screening process for new license applications to increase assurance that the material will be used as intended.

Security Rulemaking

During FY 2009, the NRC continued rulemaking activities to stabilize the security requirements that it places on its licensees. The NRC completed the rulemakings for 10 CFR Part 73, "Physical Protection of Plants and Materials," which became effective on May 26, 2009, and has a compliance date of March 31, 2010. The final rule fulfilled the Commission's intent to complete a thorough review of physical protection program requirements and orders issued after September 11, 2001, and codify them as generically applicable security requirements. Other significant additions to the security regulations include requirements for cyber security, mitigative strategies and response procedures for potential or actual aircraft attacks, and assessment and management of the interface between safety and security. The agency finalized the fitness-for-duty rule, proposed revisions to the access authorization and physical protection rule, and published a proposed rule for Nuclear Materials Management and Safeguards System database reporting. The agency also implemented interim fingerprinting requirements.

During FY 2009, the NRC continued its progress in developing regulations for security requirements for possession and use of Category 1 and 2 quantities of radioactive material. Among other provisions, the rule would include fingerprinting requirements and background investigations. The NRC posted preliminary draft rule language for public comment. The rule would include security requirements for the transportation and use of Category 1 and 2 quantities of radioactive material to provide reasonable assurance of preventing the theft or diversion of the material for malevolent purposes.

In addition, the agency made significant progress in the development of security infrastructure for new reactor licensing. The infrastructure includes the development of standard review plans for early site permits, design certifications, and COLs as well as guides for security assessment format and content. The agency also provided guidance to the industry on the new rule by developing regulatory guides for physical security, security officer training and qualification, cyber security programs, and the access authorization program. The NRC continued interactions with DHS on security infrastructure through periodic meetings. The NRC also completed its initial security review for the design certification of the General Electric ESBWR, provided technical support for a draft regulatory guide on COLs, and finished its security review of the early site permit for the Vogtle plant.

Control of Radioactive Sources

In FY 2009, the NRC continued its efforts to mitigate the potential risk of terrorist threats through enhanced security and controls for the use, storage, and transportation of byproduct material and spent nuclear fuel. In collaboration with DHS, DOE, and other Federal, State, and local agencies, the NRC continued to assess the potential use of risk-significant sources in radiological dispersal devices and to coordinate efforts to enhance radioactive source protection and security.

The NRC worked with Agreement States to implement requirements imposed on licensees that enhance the security and control of risk-significant radioactive material, including development of an inspection program to verify the implementation of these measures. In FY 2009, the NRC and Agreement States issued orders or other regulatory requirements to these licensees to require fingerprinting for those persons with unescorted access to risk-significant radioactive material. The NRC also continued activities to implement the national source tracking rule, which requires licensees to report information beginning on January 30, 2009, for inclusion in a database to track the possession of risk-significant radioactive sources. The rule requires NRC and Agreement State licensees to report transactions involving the manufacture, transfer, receipt, and disposal of nationally tracked sources (i.e., Category 1 and 2 sources from the IAEA Code of Conduct on the Safety and Security of Radioactive Sources). In response to a Government Accountability Office investigation of the ease of obtaining a new license for radioactive sources, the NRC and Agreement States have implemented a process to screen new license applications or applicants to determine, with reasonable assurance, that they will use the requested materials as intended.

The NRC continued its significant participation in implementing portions of the IAEA Code of Conduct on the Safety and Security of Radioactive Sources, as well as its participation in IAEA committees that are developing guidance documents for the security of radioactive sources during use, storage, and transport. The NRC's involvement in these committees enhances security and public safety and contributes to international and domestic regulatory consistency.

Under 10 CFR Part 110, between October 1, 2008, and June 23, 2009, the NRC issued 36 licenses for the export/import of Category 1 and 2 materials as defined by the Code of Conduct.

In FY 2009, the NRC continued efforts to establish and monitor classified information security programs for uranium enrichment vendors and a MOX facility. These efforts included readiness reviews of the Louisiana Energy Service (LES) National Enrichment Facility (NEF), in Eunice, New Mexico; USEC's Lead Cascade/American Centrifuge Project, in Piketon, Ohio; and the General Electric-Hitachi Global Laser Enrichment facility in Wilmington, North Carolina; as well as processing applications for facility security clearances under the National Industrial Security Program. These reviews included an evaluation of physical and information system security at these sites and for licensee contractors performing classified work, as well as evaluating foreign ownership, control, or influence considerations in support of the facility security clearance. In addition, NRC personnel participated in the Quadripartite Working Group and DOE meetings related to the classification and technology guides for the protection of restricted data associated with the LES project. In July 2009, NRC personnel observed the hot acceptance testing of the first two TC-12 production centrifuges assembled at the LES NEF. The NRC has also performed an acceptance review of the information security plan submitted as part of the proposed AREVA Eagle Rock Enrichment Facility at Eagle Rock, Idaho. The agency is presently working on a memorandum of understanding with DOE and the National Nuclear Security Administration on security controls for the MOX facility in Aiken, South Carolina.

Spent Fuel

In FY 2009, the NRC completed six security plan reviews for proposed independent spent fuel storage installations and issued security orders to five new independent spent fuel storage installation licensees. The security orders imposed additional security measures for physical protection, access authorization, and fingerprinting. The NRC also reviewed and approved five spent fuel transportation routes.

Costing to Goals

The NRC is working to improve its cost management capabilities to better align its costs with desired outcomes. This year's Performance and Accountability Report presents the full cost of achieving the safety and security goals for the agency's programs, Nuclear Reactor Safety and Security and Nuclear Materials Safety and Security. The cost of achieving the agency's safety goal was $992.1 million, and the cost of achieving the agency's security goal was $50.8 million (see Figure 21).

Figure 21
NRC SAFETY AND SECURITY COSTS
(In Millions)

$50.8 Security

$992.1 Safety

Organizational Excellence Objectives

The NRC has three organizational excellence objectives: openness, effectiveness, and operational excellence. These objectives are critical components of carrying out the agency's regulatory mandate to serve the American people.

Openness

The agency views nuclear regulation as the public's business and, as such, it should be transacted openly and candidly in order to maintain the public's confidence. The openness objective explicitly recognizes that the public must be informed about, and have a reasonable opportunity to participate in, the NRC's regulatory processes.

Nuclear Reactor Safety

The Office of Nuclear Reactor Regulation (NRR) maintains current documentation on the various programs of the Office of Nuclear Reactor Regulation on the office's Web site, including program processes, fact sheets, and public meeting schedules, and makes correspondence available through the Agencywide Documents Access and Management System (ADAMS). For example, the License Renewal Program portion of the office's Web site provides the public information on the licensing process, regulatory guidance, and the status of current activities associated with the renewal of licenses for commercial operating reactors. In FY 2009, the nuclear reactor regulation program exceeded its target level for ensuring that public meetings were posted on the Web site at least 10 days in advance of the meeting. The NRC conducted several successful outreach activities in the vicinity of the San Onofre and Diablo Canyon nuclear plants in California. Following these poster sessions and town hall meetings, the NRC received positive feedback from multiple stakeholders regarding the productive and effective discussion of salient topics of local interest.

Nuclear Material and Waste Safety

The NRC continues participation in the Institute of Nuclear Materials Management Spent Fuel Seminar, regional meetings of the Council of State Governments, the U.S. Transport Council meetings, and the Nuclear Energy Institute Dry Cask Storage Forum, as well as in meetings with industry and local, State, and other Federal agencies on radioactive material transportation and spent fuel storage matters.

The NRC meets with stakeholders to discuss spent fuel reprocessing issues. The staff met with NEI representatives in February 2009 to discuss NEI's white paper on a regulatory framework for reprocessing. A more comprehensive workshop that provided the perspectives of industry, intervenors, and the international community took place during the last day of the FCIX.

The NRC has also engaged stakeholders in its effort to develop a fuel cycle oversight program that has an improved degree of transparency, predictability, objectivity, and consistency, and that incorporates risk-informed and performance-based tools. Two meetings took place in June 2009 with industry representatives and members of the public, and the NRC is planning periodic public meetings as the initiative progresses.

The NRC continues to hold public meetings on issues related to fuel cycle licensing and inspections, including management and outreach meetings with State, local, and other Federal agencies for new fuel facilities under construction (LES, MOX Fuel Fabrication Facility) or under review (AREVA and General Electric-Hitachi enrichment plants).

In response to an invitation by the Native American Forum on Nuclear Issues, the NRC staff presented an overview of the process and current status of the NRC technical review of the DOE high-level waste repository license application. The staff represented the agency at a "Champions of Participation" working session that developed recommendations to President Obama's Open Government Directive. A staff member traveled to Boston and received an award as one of "CAREERS and the disABLED" magazine's Employees of the Year. The next day, staff members recruited for the NRC at the Career Expo for People with Disabilities. The NRC staff continued to hold public NRC/DOE management meetings on the license application review process for a high-level waste repository.

In the decommissioning and low-level waste arena, the NRC held public meetings in FY 2009 to discuss the decommissioning plan for the Shieldalloy materials site, the license termination plan for the Fermi power reactor, and the post shutdown decommissioning activities report for the nuclear ship Savannah. The NRC staff also met with the uranium recovery industry to discuss a draft regulatory issue summary on the NRC's draft policy for licensing in-situ recovery facilities. The NRC held 15 technical meetings with uranium recovery applicants and licensees that were open to the public, stakeholders, and Native American Tribes for observation. The NRC briefed the Commission separately on the uranium recovery program and the low-level waste program and invited representatives from other Federal and State regulatory agencies and interested stakeholders to make presentations and discuss their concerns with the Commission.

In the nuclear materials users arena, the NRC has engaged the public and stakeholder participants by providing timely notification of open meetings and by participating in national meetings of the Organization of Agreement States, the Conference of Radiation Control Program Directors, and the Health Physics Society with agency stakeholders. The NRC provides timely reviews of Freedom of Information Act requests and input to NRC administrative and Federal court proceedings. The staff completes the NRC allegations and investigations in a timely manner and provides responses to the allegers. In August 2009, 43 governor-appointed State Liaison Officers (SLO) attended the 2009 National State Liaison Officers Conference in Rockville, Maryland. The conference focused on cooperation, communication, and coordination in efforts by the NRC and the SLOs in working together to protect people and the environment. In August 2009, the NRC participated in the Bi-Annual Summit of the Yukon River Inter-Tribal Watershed Council, an organization that represents 53 Federally recognized tribes located in Alaska and 17 First Nations (Native Tribes in Canada). The NRC addressed summit attendees and familiarized them with NRC's roles, responsibilities, and its regulatory authority, and discussed staff's approach for developing Native American protocols at the NRC. The NRC also met with foreign counterparts to share information and respective lessons learned concerning the implementation of the export/import licensing provisions in the IAEA Code of Conduct on the Safety and Security of Radioactive Sources.

Effectiveness

The drive to improve performance in Government, coupled with increasing demands on the NRC's resources, requires the NRC to become more effective, efficient, and timely in its regulatory activities. The agency's initiatives related to effectiveness serve to sharpen the agency's focus on safety and security and to ensure that its available resources are optimally directed toward accomplishing the agency's mission.

Nuclear Reactor Safety

The Replace with Office of Nuclear Reactor Regulation (NRR) continues to ensure the Licensing and Oversight Programs operate effectively through the monitoring of program goals and results in accordance with agency-wide initiatives. At the program level, NRR continues to develop risk-informed and performance-based approaches to provide appropriate insights to decision makers. NRR uses state of the art methods and risk insights to improve the effectiveness and realism of NRC activities and licensing reviews resulting in high quality and timely decisions under the licensing program. NRR continues to rely and build upon industry operating experience and available information technology to improve our programs such as our efforts to optimize the ROP inspection program. NRR continues to update the infrastructure for the license renewal program, which includes updating the Generic Aging Lessons Learned (GALL) Report and Generic Environment Impact Statement (GEIS) for license renewal, to increase the efficiency of the program for present and future years.

The NRR is currently working on improving internal processes such as developing and issuing requests for additional information and safety evaluations in support of license amendment requests. NRR is also revising scheduling procedures for LARs to be more consistent with the licensee's need for and the complexity of the LAR. NRR is also examining the feasibility of expanding enterprise project management from selected applications to all licensing actions. This project is projected to be used as a replacement for the current legacy scheduling system in use at NRR, which will result in better project management.

Overall, the NRR programs continue to be effective through use of lessons learned from event responses, inspections and operational experience, and feedback leading to improved inspection and licensing programs in FY 2009.

Nuclear Material and Waste Safety

The NRC began a number of initiatives to improve the efficiency and effectiveness of its radioactive material transportation and spent fuel storage program. Examples include (1) working cooperatively with industry to develop procedures for license application acceptance reviews, (2) documenting a transparent process to prioritize new and ongoing activities, and (3) developing an approach for performing more focused regulatory reviews of calculations and methodologies using risk information, lessons learned, and operating experience to guide the depth of the reviews.

In 2009 the agency maintained the viability of the High-Level Waste Repository Program and the Center for Nuclear Waste Regulatory Analysis. The NRC met its review schedule and provided hearing support for the review of 319 contentions.

The NRC made substantial progress in updating the fuel facility regulatory infrastructure (regulatory guides and standard review plan) to reflect current agency positions and operating experience. This ensures that regulatory actions are more consistent, predictable, and transparent.

In the decommissioning and low-level waste arena, as part of the NRC's license application review process, staff performs an acceptance review to determine if the license application contains adequate information to begin a detailed technical review. This ensures that the NRC staff does not expend resources reviewing submissions that contain incomplete or inadequate information. Further, to assist in the review of in situ recovery applications, the NRC published a final GEIS. The GEIS contributes to the agency's application review process by addressing common environmental issues associated with the construction, operation, and decommissioning of in situ recovery facilities, as well as the ground water restoration at such facilities, if they are located in particular regions of the western United States. The NRC estimates that issuance of the GEIS will result in a total savings for all application reviews of as much as $7 million and will reduce review time by 2 years per application. Additionally, the NRC has initiated negotiations with the Bureau of Land Management to develop a memorandum of understanding to allow for cooperation between the two agencies on environmental review documents to meet the requirements of the National Environmental Policy Act.

Operational Excellence

The agency strives for operational excellence in carrying out all of its regulatory responsibilities. This objective focuses on activities relating to the management of finances, human capital, information, and infrastructure. This objective supports the agency by ensuring that the necessary infrastructure is in place to accomplish the agency's mission.

Financial Management

The Office of the Chief Financial Officer (OCFO) made substantial progress in its effort to modernize the agency's financial systems during FY 2009. It implemented an e-Travel system across the agency, upgraded the Budget Formulation System, initiated the upgrade of the Time and Labor System to a Web-based paperless system, and is proceeding with a systems modernization effort for the agency's core accounting system.

OCFO improved outreach by holding a public meeting to discuss and inform stakeholders of the agency's fee and budget processes. It expanded a cross-servicing agreement with the agency's shared service provider (U.S. Department of the Interior/National Business Center) to include processing agency-wide obligations.

OCFO continued to achieve operational excellence in financial reporting. The agency received an unqualified opinion on the FY 2008 financial statement with no material weaknesses and its eighth consecutive Certificate of Excellence in Accountability Reporting award. The Mercatus Center also credited the NRC-OCFO with a high ranking for the agency's Performance and Accountability Report. The agency strengthened its risk assessment and reasonable assurance process to be a more thoughtful, deliberative process by focusing on management support, updating guidance and processes, and providing online training to all agency employees.

OCFO completed an assessment of its organizational functions, processes, and roles and responsibilities. Additionally, the agency identified Federal Chief Financial Officer best practices and compared them to those used by the NRC's OCFO. The NRC will execute an action plan to align the agency's OCFO to the Federal best practices. Operationally, OCFO placed increased emphasis on improving execution of the NRC budget to align the spending forecast to the agency's acquisition planning. OCFO continues to lead an agency-wide effort to significantly reduce prior fiscal year unliquidated obligations. It also emphasized better customer service and outreach to internal and external NRC stakeholders.

Management of Human Capital

There has been and continues to be a critical shortage of personnel in the nuclear sector as the current workforce retires and normal attrition occurs. The NRC has an ongoing education grant program to provide grants to educational institutions in the areas of curriculum development, faculty development, fellowships, scholarships to 4-year institutions, and scholarships to trade schools and community colleges. These grants assist in the development of the next-generation nuclear workforce. The NRC made more than 85 grants in FY 2008 to educational institutions and an additional 100 or more grants in FY 2009. These grants focus on the areas of nuclear engineering, health physics, radiochemistry, and other related areas that benefit the nuclear sector.

In FY 2009, the NRC implemented several of the recommendations developed by a Lean Six Sigma process review team to meet the timeliness standards established by the U.S. Office of Personnel Management end-to-end hiring model. This effort will reduce the time necessary to bring personnel into the agency.

The agency launched the NRC Knowledge Center, an agency-wide collection of electronic Communities of Practice designed to enable staff to collaborate, capture, and share knowledge in order to build organizational memory. The NRC established an Expertise Exchange to capture the lessons learned and best practices from the agency's most experienced staff. The NRC is also contacting experts in knowledge management and strategic workforce planning across the Federal government and in industry to identify best practices and lessons learned.

Information Technology and Information Management

The NRC has engaged internal stakeholders in efforts to identify impediments to program performance and information availability that can be addressed with information technology. As a result of these activities, the agency identified three major focus areas for improving business operations through the application of information technology and information management: working from anywhere, organizational productivity, and universal access.

The top priority area, working from anywhere, refers to enabling the staff to communicate securely and use the needed systems and information, whether the person is telecommuting, on travel, or moving between the NRC's various locations. Accomplishments in this area included increasing the availability of wireless handheld devices and secure laptops, instituting a "loaner laptop" program, upgrading the agency's video-teleconferencing system, modernizing the agency's remote access system used by telecommuters and the agency's resident inspectors, increasing the capacity of the agency's Internet connection to enable more effective Web streaming of agency meetings, increasing electronic access to industry codes and standards needed by the licensing staff, and providing electronic access to the agency's technical and law library catalogs.

Improving organizational productivity means enabling individuals and groups to work more efficiently to accomplish the agency's mission. Progress in this area included making collaboration software available agencywide and enabling Web-based electronic meetings.

The ultimate goal of universal access is to enable authorized individuals to access NRC facilities, information technology systems, and needed information securely through a single access authorization mechanism. Progress on universal access included improvements to the NRC's network directory necessary to lay the groundwork for reducing the number of required sign-ons, elimination of duplicate sign-on to the NRC network, and successful implementation of managed public key infrastructure for external and internal use.

In the area of records management, the NRC made improvements in the automation of e-mail, document capture, and records retention and disposition. Knowledge management continues to be a key component of organizational excellence to ensure that organizational knowledge is retained.

Infrastructure Management

This year, the NRC Office of Administration (ADM) completed the acquisition of additional interim buildings near the White Flint Headquarters campus to address the growth in staff to support the New Reactor Program and other licensing activities. The NRC has a total of four interim locations near its Headquarters. Pending completion of a final building to bring all staff back to the Headquarters campus, ADM established a Staying Connected Working Group to develop strategies to maintain the feeling of employee cohesiveness. ADM has taken several steps to improve the efficiency of support services and make it easier to accomplish agency goals. It has expanded shuttle services to transport employees who need to attend meetings, special events, and training in various Headquarters-area buildings. It has also created work stations in each Headquarters-area location equipped with telephones and computers to enable employees to conduct business while at locations other than their primary duty station. In addition, ADM has increased physical security to meet the needs of the expanding number of facilities.

In addition to subjecting all NRC employees to both preassignment and random drug testing, ADM has now implemented a similar drug testing program for NRC contractors. The new program provides both preassignment and random drug testing for badged and unbadged NRC contractors who are in sensitive positions (i.e., those operating government vehicles and carrying weapons, as well as those who require unescorted access to nuclear power plants and access to safeguards or classified information) or those who admit to recent illegal drug use. The expansion to include this program increases the effectiveness of personnel security by ensuring that those contractors in sensitive positions conform to the NRC's Drug-Free Workplace plan.

Program Evaluations

The NRC's Strategic Plan for Fiscal Years 2008–2013 describes a number of ongoing program evaluations that the NRC was scheduled to conduct for a self-assessment of its regulatory operations. This section lists the results of these program evaluations.

Operator Licensing Program

An NRC review team evaluated the overall effectiveness of the Region I and IV operator licensing programs and their adherence to the guidance contained in NUREG-1021, "Operator Licensing Examination Standards for Power Reactors," and other policy documents. The review divided the operator licensing programs into seven functional areas, rated as "satisfactory" or "needs improvement." Overall, the operator licensing programs in Regions I and IV are being conducted in accordance with NUREG-1021. For both regions, the review team assessed all areas as satisfactory. The review team also commended the regions' efforts to improve the quality of their examination packages on ADAMS.

Reactor Oversight Program

The NRC completed the calendar year 2008 ROP self-assessment in April 2009. The report, SECY-09-0054, "Reactor Oversight Process Self-Assessment for Calendar Year 2008," dated April 6, 2009, is available through the NRC public Web site.

The results of this self-assessment indicate that the ROP met its program goals and achieved its intended outcomes. The staff found the ROP objective to be risk informed, understandable, and predictable, and the ROP met the agency goals of ensuring safety, openness, and effectiveness, as listed in the NRC's Strategic Plan Fiscal Years 2008–2013. The NRC staff maintained its focus on stakeholder involvement and continued to improve various aspects of the ROP. The staff implemented several ROP improvements in calendar year 2008 to address issues raised by the Commission, recommended by independent reviews, and obtained from internal and external stakeholder feedback.

The NRC inspection and assessment program independently verified that nuclear power plants were operated safely and securely. The staff revised the assessment program to incorporate lessons learned from the implementation of the safety culture enhancements and continued to ensure that the NRC staff and licensees acted as necessary to address identified performance issues. The staff continues to improve the performance indicator program to ensure that the performance indicators are meaningful inputs to the ROP. It actively solicits input from the NRC's internal and external stakeholders to further improve the ROP based on stakeholder feedback and lessons learned.

Integrated Materials Performance Evaluation Program Reviews of Selected NRC Regional Offices

The NRC, with the assistance of the Agreement States, completed eight Integrated Materials Performance Evaluation Program reviews to determine the adequacy and compatibility of the programs in evaluated Agreement States and one review for the materials licensing and inspection program and uranium recovery inspection program in NRC Region IV. Region IV was found satisfactory (the highest level) for all areas of the review, and there were no recommendations for the region.

Fuel Cycle Licensing and Inspection Program

The NRC's Fuel Cycle Licensing and Inspection Program is the agency's program to regulate the nation's non-defense related fuel fabrication facilities. Its licensing program is designed to issue licenses to facilities to receive title to, own, acquire, deliver, receive, possess, use, and transfer special nuclear material (SNM). Further, this program is necessary to verify that companies can safely use SNM prior to taking possession and starting operations. The inspection program's purpose is to obtain objective information that will permit the NRC to assess whether licensees operate licensed fuel cycle facilities

safely and in compliance with regulations, and that licensee activities do not pose undue safety and safeguards risks. This inspection program needs to be performed routinely since companies continue to make changes to facilities, staff, and operations.

The contract was awarded in September 2009. A kickoff meeting was conducted with the contractor on October 20, 2009.

The objective of this contract is to obtain expert analysis from a qualified entity who is familiar with the PART review process to assist NRC staff by performing a program assessment and gap analysis, and by developing recommendations to strengthen program performance.

Data Sources and Quality

The NRC's data collection and analysis methods are driven largely by its regulatory mandate. Specifically, the NRC's mission is to regulate the Nation's civilian use of byproduct, source, and special nuclear materials to ensure adequate protection of public health and safety, promote the common defense and security, and protect the environment. In undertaking this mission, the NRC oversees nuclear power plants, nonpower reactors, nuclear fuel facilities, interim spent fuel storage, radioactive material transportation, disposal of nuclear waste, and the industrial and medical uses of nuclear materials. Section 208 of the Energy Reorganization Act of 1974, as amended, requires the NRC to inform Congress of incidents or events that the Commission determines to be significant from the standpoint of public health and safety. The NRC developed the abnormal occurrence criteria to comply with the legislative intent of the Energy Reorganization Act to determine which events should be considered significant. Based on those criteria, the NRC prepares an annual "Report to Congress on Abnormal Occurrences" (NUREG-0090, Volume 31), which is available on the agency's public Web site at http://www.nrc.gov/reading-rm/doc-collections/nuregs/staff/sr0090.

One important characteristic of this report is that the data presented normally originate from external sources such as Agreement States and NRC licensees. The NRC finds these data credible because: (1) agency regulations require Agreement States, licensees, and other external sources to report the necessary information; (2) the NRC maintains an aggressive inspection program that, among other activities, includes auditing licensee programs and evaluating Agreement State programs to ensure that they are reporting the necessary information as required by the agency's regulations; and (3) the agency has established procedures for inspecting and evaluating licensees. The NRC employs multiple database systems to support this process, including the Licensee Event Report Search System, the Accident Sequence Precursor Database, the Nuclear Materials Events Database, and the Radiation Exposure Information Report System. In addition, nonsensitive reports submitted by Agreement States and NRC licensees are available to the public through the NRC's ADAMS, accessible through the agency's public Web site at http://www.nrc.gov/reading-rm/adams.html.

As stated above, the NRC has established procedures for the systematic review and evaluation of events reported by both NRC and Agreement State licensees. The NRC's objective is to identify events that are significant from the standpoint of public health and safety based on criteria that include specific thresholds. The NRC verifies the reliability and technical accuracy of event information reported to the agency. The NRC periodically inspects licensees and reviews Agreement State programs. In addition, NRC Headquarters, the regional offices, and Agreement States hold periodic conference calls to discuss event information. Events identified as meeting the abnormal occurrence criteria are validated and verified before being reported to Congress.

Information Security

The agency's information security (IS) program: (1) protects NRC and licensee information and information systems from unauthorized access, use, disclosure, disruption, modification, or destruction; (2) protects electronic control functions from unauthorized access or manipulation; and (3) ensures that adequate controls for protecting security-related information are used in the conduct of NRC business, both internal and external to the agency. The NRC information security program includes measures to accomplish the following:

1. Ensure that IS requirements, standards, and guidance are clear, concise, appropriate, and able to mitigate potential adverse effects if sensitive information is compromised.

2. Ensure that security controls for information owned by, or under the control of the NRC, are consistent with established IS controls; that security controls for information are operating as intended and that they are having the desired impact; and that similar controls for licensees regulated by the NRC are in compliance with NRC IS regulations.

3. Ensure that the NRC evaluate suspected or actual IS violations and consider appropriate sanctions.

4. Ensure that the NRC has made sufficient preparations for IS-related emergencies and incidents.

5. Ensure internal IS program components complement each other and are periodically evaluated and improved.

Performance Data Completeness and Reliability

In order to manage for results, it is essential for the NRC to assess the completeness and reliability of NRC performance data. Comparisons of actual performance with the projected levels are possible only if the data used to measure performance are complete and reliable. Consequently, the Reports Consolidation Act of 2000 requires the Chairman of the NRC to assess the completeness and reliability of the performance data used in the Report to Congress. The process for ensuring that the data are complete and reliable requires offices to complete a template for submission to the Chief Financial Officer for every performance measure certifying that the applicable Office Director has approved the data submitted.

Data Completeness

The NRC considers data to be complete if the agency reports actual performance data for every performance goal and indicator in the annual plan. Actual performance data include preliminary data if those are the only data available when the agency sends its report to the President and Congress. The NRC has reported actual or preliminary data for every strategic

and performance goal measure; consequently, the data presented in this report meet these requirements for data completeness.

Data Reliability

The NRC considers data to be reliable when agency managers and decision makers use the data in carrying out their regulatory responsibilities. The data presented in this report meet this requirement for data reliability because NRC managers and senior leaders regularly use the reported data in the course of their duties.

Annual NRC-sponsored Regulatory Information Conference (RIC) held March 10-12, 2009, in Bethesda, MD.

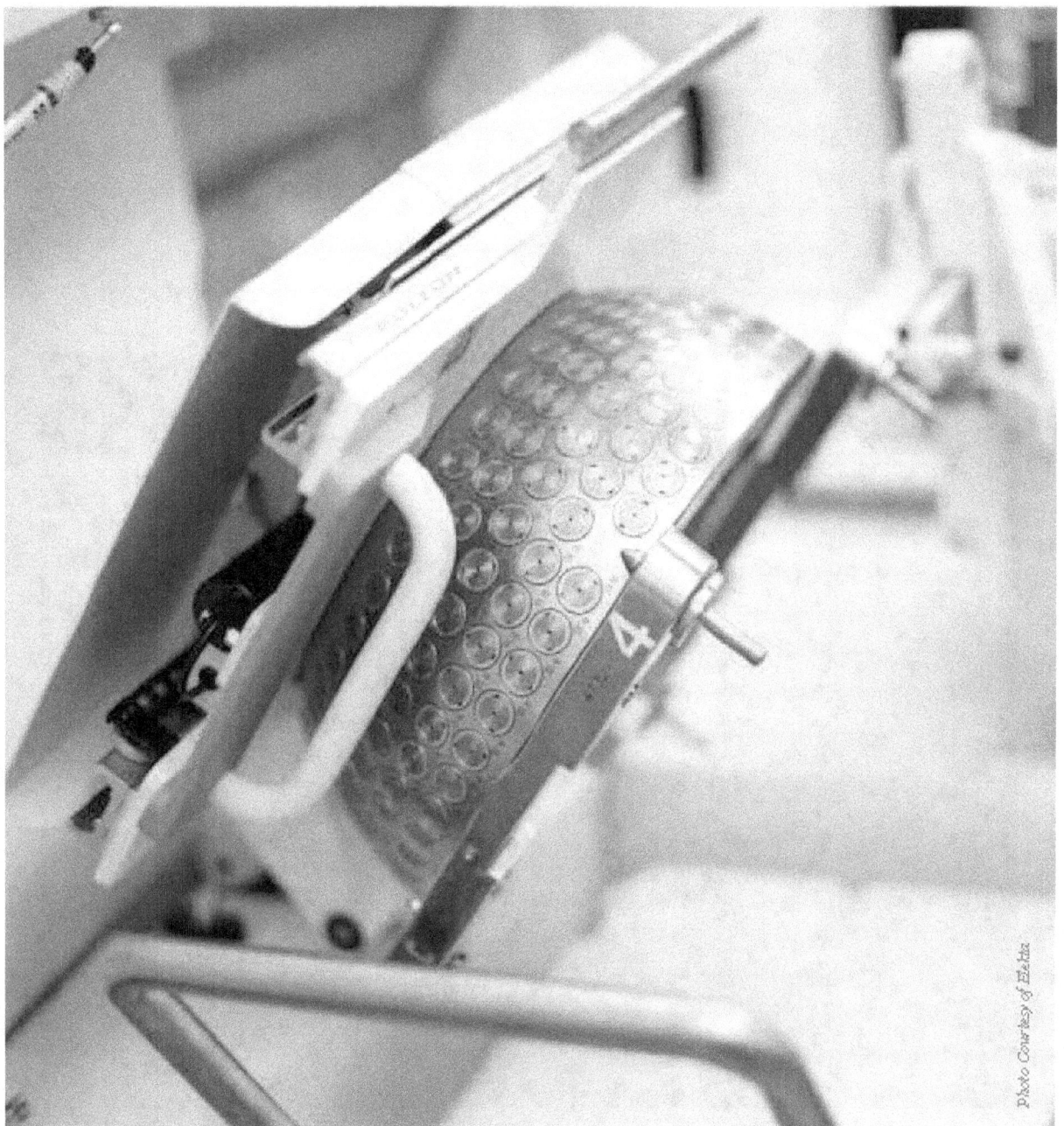

Photo Courtesy of Elekta

The Gamma Knife utilizes a technique called stereotactic radiosurgery, which uses multiple beams of radiation converging in three dimensions to focus precisely on a small volume, such as a tumor, permitting intense doses of radiation to be delivered to that volume safely.

Photo Courtesy of NRC Photo Library

NRC Region IV inspector Bob Evans performing an inspection at Salmon River Uranium Development Corp., an abandoned mill located near North Fork, ID.

Chapter 3

Financial Statements and Auditor's Report

Photo Courtesy of NRC Photo Library

South Texas Project nuclear power plant is located near Bay City, TX. It is operated by South Texan Project Nuclear Operating Company.

Photo Courtesy of NRC Photo Library

Resident inspector staff at Palo Verde Nuclear Generating Station near Wintersburg, AZ standing in front of the independent spent fuel storage installation pad. From left to right: Joseph Bashore, Ryan Treadway, Michelle Catts, and Jim Melfi

A Message from the Chief Financial Officer

I am pleased to present the financial statements for the U.S. Nuclear Regulatory Commission (NRC) Fiscal Year (FY) 2009 Performance and Accountability Report. For the sixth consecutive year, an independent auditor has rendered an unqualified opinion on the NRC financial statements. During FY 2009, the NRC successfully enhanced its procedures for estimating accrued accounts payable and eliminated the remaining significant deficiency from our prior year audits. Additionally, for the fourth consecutive year, no material weaknesses were identified during NRC testing to meet the requirements of Office of Management and Budget Circular A-123, Appendix A, "Internal Controls Over Financial Reporting."

This past year, the NRC increased its focus on modernizing financial systems, improving internal controls, and enhancing financial planning and performance. Examples of our FY 2009 accomplishments include the following:

- Implementing a new structure for the FY 2011 budget to better align it with the agency mission, improve the transparency of budget requests, and facilitate improved costing of regulatory products and support services.
- Enhancing agency budget execution to improve FY 2009 funds utilization and recover $28 million of unused prior year funds from completed contracts.
- Modernizing our financial systems by implementing a new eTravel system and upgrading our Web-based budget formulation system to increase its capabilities and reliability.
- Completing the Federal Information Security Management Act Certification and Accreditation for the License Fee Billing System which brought the NRC into substantial compliance with the Federal Financial Management Improvement Act of 1996.
- Redesigning the agencywide risk assessment process to support the NRC internal control program and implementing an online internal control training module.
- Creating an agencywide integrated project team, selecting a vendor, and documenting system requirements to prepare for the transition to a new core financial system at the start of FY 2011.

Our progress in FY 2009 puts the NRC in a good position to address FY 2010 challenges as we continue to improve our financial systems and processes. The NRC will transition five stand-alone legacy financial system functions into a new core financial system that will interface with nine remaining financial and program management systems. The NRC Time and Labor Reporting System will also be updated to a Web-based version to support the core financial system transition. The NRC is also preparing to meet the anticipated challenges of supporting ongoing financial operations while simultaneously supporting the testing and startup of the new core financial system. The new core financial system will play an integral role in meeting our goal of making the NRC a more transparent, efficient, and effective organization.

The NRC is committed to ensuring the safety and security of the Nation's civilian use of nuclear materials in the most effective and efficient manner. Over the past few years, the NRC has experienced unprecedented growth in its budget to regulate the Nation's expanding nuclear industry. Our continued excellent financial performance during this period of significant budget growth is a tribute to the careful stewardship of taxpayer resources by the NRC staff. I am proud of the progress we have made in the past year to promote sound business practices in the conduct of our regulatory mission and am confident that the NRC will continue to make future improvements.

J.E. Dyer
Chief Financial Officer
November 13, 2009

Principal Statements

BALANCE SHEET
(In Thousands)

As of September 30,	2009	2008
Assets		
Intragovernmental		
Fund balance with Treasury (Note 2)	$ 448,632	$ 393,478
Accounts receivable (Note 3)	4,907	4,692
Other-Advances and prepayments	3,340	4,121
Total intragovernmental	456,879	402,291
Accounts receivable, net (Note 3)	123,217	116,684
Property and equipment, net (Note 4)	31,624	35,475
Other	32	28
Total Assets	$ 611,752	$ 554,478
Liabilities		
Intragovernmental		
Accounts payable	$ 13,977	$ 12,360
Other (Note 5)	5,489	4,844
Total intragovernmental	19,466	17,204
Accounts payable	37,023	41,763
Federal employee benefits (Note 6)	7,628	7,059
Other (Note 5)	80,639	70,948
Total Liabilities	144,756	136,974
Net Position		
Unexpended appropriations	338,637	289,269
Cumulative results of operations (Note 8)	128,359	128,235
Total Net Position	466,996	417,504
Total Liabilities and Net Position	$ 611,752	$ 554,478

The accompanying notes to the principal statements are an integral part of this statement.

STATEMENT OF NET COST
(In Thousands)

For the years ended September 30,	2009	2008
Nuclear Reactor Safety and Security		
Gross costs	$ 796,898	$ 705,832
Less: Earned revenue	(794,007)	(725,840)
Total Net Cost of Nuclear Reactor Safety and Security (Note 9)	2,891	(20,008)
Nuclear Materials and Waste Safety and Security		
Gross costs	245,961	238,219
Less: Earned revenue	(78,460)	(71,740)
Total Net Cost of Nuclear Materials and Waste Safety and Security (Note 9)	167,501	166,479
Net Cost of Operations	$ 170,392	$ 146,471

The accompanying notes to the principal statements are an integral part of this statement.

STATEMENT OF CHANGES IN NET POSITION
(In Thousands)

For the years ended September 30,	2009	2008
Cumulative Results of Operations		
Beginning Balance	$ 128,235	$ 27,164
Budgetary Financing Sources		
Appropriations used	89,309	98,172
Non-exchange revenue (Note 11)	-	-
Transfers-in/out without reimbursement	49,000	29,025
Other Financing Sources		
Imputed financing from costs absorbed by others (Note 11)	32,207	26,911
Other (Note 16)	-	93,434
Total Financing Sources	170,516	247,542
Net Cost of Operations	(170,392)	(146,471)
Net Change	124	101,071
Cumulative Results of Operations	$ 128,359	$ 128,235
Unexpended Appropriations		
Beginning Balance	$ 289,269	$ 254,027
Budgetary Financing Sources		
Appropriations received	138,677	133,414
Appropriations used	(89,309)	(98,172)
Total Budgetary Financing Sources	49,368	35,242
Total Unexpended Appropriations	338,637	289,269
Net Position	$ 466,996	$ 417,504

The accompanying notes to the principal statements are an integral part of this statement.

STATEMENT OF BUDGETARY RESOURCES
(In Thousands)

For the years ended September 30,	2009	2008
Budgetary Resources		
Unobligated balance, brought forward, October 1	$ 78,990	$ 72,160
Recoveries of prior year unpaid obligations		
Actual	28,371	21,937
Budget authority		
Appropriation	1,045,517	926,074
Spending authority from offsetting collections		
Reimbursements earned-collected	8,429	6,709
Reimbursements earned-change in receivables	375	222
Change in unfilled customer orders-advance received	333	1,645
Change in unfilled customer orders-without advance	3,190	65
Subtotal-spending authority from offsetting collections	12,327	8,641
Total Budgetary Resources	**$ 1,165,205**	$ 1,028,812
Status of Budgetary Resources		
Obligations incurred (Note 12)		
Direct	$ 1,073,782	$ 941,942
Reimbursable	10,297	7,880
Subtotal	1,084,079	949,822
Unobligated balance		
Apportioned	66,699	69,024
Exempt from apportionment	7,609	9,853
Subtotal	74,308	78,877
Unobligated balance, not available	6,818	113
Total Status of Budgetary Resources	**$ 1,165,205**	$ 1,028,812
Change in Obligated Balance		
Obligated balance, net		
Unpaid obligations brought forward, October 1	$ 314,488	$ 270,894
Obligations incurred, net	1,084,079	949,822
Gross outlays	(999,133)	(884,004)
Recoveries of prior year unpaid obligations, actual	(28,371)	(21,937)
Change in uncollected customer payments, from Federal sources	(3,565)	(287)
Obligated balance, net, end of period		
Unpaid obligations	375,201	318,626
Uncollected customer payments, from Federal sources	(7,703)	(4,138)
Total unpaid obligated balance, net, end of period	$ 367,498	$ 314,488
Net outlays		
Gross outlays	$ 999,133	$ 884,004
Offsetting collections	(8,762)	(8,354)
Distributed offsetting receipts	(857,839)	(763,640)
Net Outlays	**$ 132,532**	$ 112,010

The accompanying notes to the principal statements are an integral part of this statement.

Notes to the Principal Statements

(All Tables are Presented in Thousands)

Note 1.
SUMMARY OF SIGNIFICANT ACCOUNTING POLICIES

A. Reporting Entity

The U.S. Nuclear Regulatory Commission (NRC) is an independent regulatory agency of the Federal Government that was created by the U.S. Congress to regulate the Nation's civilian use of byproduct, source, and special nuclear materials to ensure adequate protection of the public health and safety, to promote the common defense and security, and to protect the environment. Its purposes are defined by the Energy Reorganization Act of 1974, as amended, along with the Atomic Energy Act of 1954, as amended, which provide the foundation for regulating the Nation's civilian use of nuclear materials.

The NRC operates through the execution of its congressionally approved appropriations for Salaries and Expenses and the Office of the Inspector General, including funds derived from the Nuclear Waste Fund. In addition, the U.S. Agency for International Development (USAID) provides transfer appropriations to develop nuclear safety, regulatory authorities, and independent oversight of nuclear reactors in Russia, Ukraine, Kazakhstan, Georgia, and Armenia.

B. Basis of Presentation

These principal statements report the financial position and results of operations of the NRC as required by the Chief Financial Officers Act of 1990 and the Government Management Reform Act of 1994. These financial statements were prepared from the books and records of the NRC in conformance with generally accepted accounting principles (GAAP) of the United States and the form and content for entity financial statements specified by the Office of Management and Budget (OMB) in Circular No. A-136, "Financial Reporting Requirements."

GAAP for Federal entities are the standards prescribed by the Federal Accounting Standards Advisory Board, which is the official body for setting the accounting standards of the U.S. Government. These statements are, therefore, different from the financial reports, also prepared by the NRC pursuant to OMB directives, which are used to monitor and control the NRC's use of budgetary resources.

The NRC has not presented a Statement of Custodial Activity because the amounts involved are immaterial and incidental to its operations and mission.

C. Budgets and Budgetary Accounting

Budgetary accounting measures appropriation and consumption of budget spending authority or other budgetary resources and facilitates compliance with legal constraints and controls over the use of Federal funds. Under budgetary reporting principles, budgetary resources are consumed at the time of purchase. Assets and liabilities, which do not consume current budgetary resources, are not reported, and only those liabilities for which valid obligations have been established are considered to consume budgetary resources.

For the past 35 years, Congress has enacted no-year appropriations, which are available for obligation by the NRC until expended. For FY 2009, the Omnibus Appropriations Act, 2009 requires the NRC to recover approximately 90 percent of its new budget authority by assessing fees for licensing and inspection activities.

D. Basis of Accounting

These financial statements reflect both accrual and budgetary accounting transactions. Under the accrual method, revenues are recognized when earned and expenses are recognized when a liability is incurred, without regard to receipt or payment of cash. Budgetary accounting is also used to record the obligation of funds prior to the accrual-based transaction. The Statement of Budgetary Resources presents budgetary resources available to the NRC and changes in obligations during the year. Interest on borrowings of the U.S. Department of the Treasury (Treasury) is not included as a cost to the NRC programs and is not included in the accompanying financial statements.

E. Revenues and Other Financing Sources

The NRC is required to offset its appropriations by revenue received during the fiscal year from the assessment of fees. The NRC assesses two types of fees to recover its budget authority: (1) fees assessed under Title 10 of the *Code of Federal Regulations* (10 CFR) Part 170, "Fees for Facilities, Materials, Import and Export Licenses, and Other Regulatory Services under the Atomic Energy Act of 1954, as Amended," for licensing, inspection, and other services under the authority of the Independent Offices Appropriation Act of 1952 to recover the NRC's costs of providing individually identifiable services to specific applicants and licensees; and (2) annual fees assessed for nuclear facilities and materials licensees under 10 CFR Part 171, "Annual Fees for Reactor Licenses and Fuel Cycle Licenses and Material Licenses." Licensing revenues are recognized on a straight-line basis over the licensing period. Inspection fees are recorded as revenues when the services are performed.

For accounting purposes, appropriations are recognized as financing sources (appropriations used) at the time goods and services are received. At the end of the fiscal year, appropriations recognized are reduced by the amount of assessed fees collected during the fiscal year to the extent of new budget authority for the year. Collections which exceed the new budget authority are held to offset subsequent years' appropriations. Appropriations expended for property and equipment are recognized as expenses when the asset is consumed in operations as reflected by depreciation and amortization expense.

F. Fund Balance with Treasury

The NRC's cash receipts and disbursements are processed by the Treasury. The Fund Balance with Treasury is primarily appropriated funds that are available to pay current liabilities and to finance authorized purchase commitments. Fund Balance with Treasury represents the NRC's right to draw on the Treasury for allowable expenditures.

G. Accounts Receivable

Accounts receivable consist of amounts owed to the NRC by other Federal agencies and the public. Amounts due from the public are presented net of an allowance for uncollectible accounts. The allowance is determined based on the age of the receivable and allowance rates established from historical experience. Receivables from Federal agencies are expected to be collected; therefore, there is no allowance for uncollectible accounts for Federal agencies.

H. Non-Entity Assets

Non-entity assets consist of miscellaneous penalties and interest due from the public, which, when collected, must be transferred to the Treasury.

I. Property and Equipment

Property and equipment consist primarily of typical office furnishings, leasehold improvements, nuclear reactor simulators, and computer hardware and software. The costs of internal use software include the full cost of salaries and benefits for agency personnel involved in software development. The NRC has no real property. The land and buildings in which the NRC operates are provided by the General Services Administration (GSA), which charges the NRC rent that approximates the commercial rental rates for similar properties.

Property with a cost of $50 thousand or more per unit and a useful life of 2 years or more is capitalized at cost and depreciated using the straight-line method over the useful life. Other property items are expensed when purchased. Normal repairs and maintenance are charged to expense as incurred.

J. Accounts Payable

The NRC uses an estimation methodology to calculate the accounts payable balance which represents costs for billed and unbilled goods and services received (prior to year end) that are unpaid. The NRC calculates the accounts payable estimate by analyzing the actual activity for a sample of open obligations. From this analysis, an algorithm is developed to estimate the accounts payable balance.

K. Liabilities Not Covered by Budgetary Resources

Liabilities represent the amount of monies or other resources that are likely to be paid by the NRC as the result of a transaction or event that has already occurred. No liability can be paid by the NRC absent an appropriation. Liabilities for which an appropriation has not been enacted are classified as "Liabilities Not Covered by Budgetary Resources." Also, the NRC liabilities arising from sources other than contracts can be abrogated by the Government acting in its sovereign capacity.

Intragovernmental

The NRC records a liability to the U.S. Department of Labor (DOL) for Federal Employees Compensation Act (FECA) benefits paid by DOL on behalf of the NRC.

Federal Employee Benefits

Federal employee benefits represent the actuarial liability for estimated future FECA disability benefits. The future workers' compensation estimate was generated by DOL from an application of actuarial procedures developed to estimate the liability for FECA, which includes the expected liability for death, disability, medical, and miscellaneous costs for approved compensation cases. The liability is calculated using historical benefit payment patterns related to a specific incurred period to predict the ultimate payments related to that period. These projected annual benefit payments are discounted to present value. The interest rate assumptions utilized for discounting benefits are 4.22 percent and 4.37 percent for FY 2009 and FY 2008, respectively.

Other

Accrued annual leave represents the amount of annual leave earned by NRC employees but not yet taken.

L. Contingencies

Contingent liabilities are those for which the existence or amount of the liability cannot be determined with certainty pending the outcome of future events. The NRC is a party to various administrative proceedings, legal actions, environmental suits, and claims brought by or against it. Based on the advice of legal counsel concerning contingencies, it is the opinion of management that the ultimate resolution of these proceedings, actions, suits, and claims will not materially affect the agency's financial statements. As of September 30, 2009 and 2008, the NRC was not a party to a case in which an adverse outcome was probable or reasonably possible.

M. Annual, Sick, and Other Leave

Annual leave is accrued as it is earned and the accrual is reduced as leave is taken. Each year, the balance in the accrued annual leave liability account is adjusted to reflect current pay rates. To the extent that current or prior year funding is not available to cover annual leave earned but not taken, funding will be obtained from future financing sources. Sick leave and other types of nonvested leave are expensed as taken.

N. Retirement Plans

The NRC employees belong to either the Federal Employees Retirement System (FERS) or the Civil Service Retirement System (CSRS). For FY 2009 and FY 2008, for employees belonging to FERS, the NRC withheld 0.8 percent of base pay earnings, in addition to Federal Insurance Contribution Act (FICA) withholdings, and matched the withholdings with an 11.2 percent contribution. The sum is transferred to the Federal Employees Retirement Fund. For employees covered by CSRS, the NRC withholds 7 percent of base pay earnings. The NRC matched this withholding with a 7 percent contribution in FY 2009 and FY 2008.

The Thrift Savings Plan (TSP) is a retirement savings and investment plan for employees belonging to either FERS or CSRS. The maximum percentage of base pay that an employee participating in FERS or CSRS may contribute is unlimited in 2009 and 2008, subject to the maximum contribution of $16.5 thousand in 2009 and $15.5 thousand in 2008. For employees participating in FERS, the NRC automatically contributes 1 percent of base pay to their account and matches contributions up to an additional 4 percent. For employees participating in CSRS, there is no NRC matching of the contribution. The sum of the employees' and NRC's contributions are transferred to the Federal Retirement Thrift Investment Board.

The NRC does not report on its financial statements FERS and CSRS assets, accumulated plan benefits, or unfunded liabilities, if any, applicable to its employees. Reporting such amounts is the responsibility of the U.S. Office of Personnel Management. The portion of the current and estimated future outlays for CSRS not paid by the NRC is included in NRC's financial statements as an imputed financing source in NRC's Statement of Changes in Net Position and as program costs on the Statement of Net Cost.

O. Leases

The NRC's capital leases are for personal property consisting of reproduction equipment which is installed at NRC headquarters. For FY 2009, there are eight capital leases with terms of 5 years, consisting of two capital leases added in FY 2008 with an interest rate of 3.99 percent, two capital leases that were added in FY 2007 with an interest rate of 4.58 percent, one capital lease in FY 2006 with an interest rate of 4.25 percent, and three capital leases for FY 2005 with an interest rate of 4.13 percent. The reproduction equipment is depreciated over 5 years using the straight-line method with no salvage value.

Operating leases consist of real property leases with GSA. The leases are for NRC's headquarters and regional offices. The GSA charges the NRC lease rates which approximate commercial rates for comparable space.

P. Pricing Policy

The NRC provides nuclear reactor and materials licensing and inspection services to the public and other Government entities. In accordance with OMB Circular No. A-25, "User Charges," and the Independent Offices Appropriation Act of 1952, the NRC assesses fees under 10 CFR Part 170 for licensing and inspection activities to recover the full cost of providing individually identifiable services.

The NRC's policy is to recover the full cost of goods and services provided to other Government entities where (1) the services performed are not part of its statutory mission and (2) the NRC has not received appropriations for those services. Fees for reimbursable work are assessed at the 10 CFR Part 170 rate with minor exceptions for programs that are nominal activities of the NRC.

Q. Net Position

The NRC's net position consists of unexpended appropriations and cumulative results of operations. Unexpended appropriations represent appropriated spending authority that is unobligated and has not been withdrawn by the Treasury and obligations that have not been paid. Cumulative results of operations represent the excess of financing sources over expenses since inception.

R. Use of Management Estimates

The preparation of the accompanying financial statements in accordance with generally accepted accounting principles requires management to make certain estimates and assumptions that affect the reported amounts of assets, liabilities, revenues, and expenses. Actual results could differ from those estimates.

S. Appropriation Transfers

The NRC is a party to allocation transfers with the U.S. Agency for International Development (USAID) as a receiving (child) entity. These transfers are for the international development of nuclear safety and regulatory authorities in Russia, Ukraine, Kazakhstan, Georgia and Armenia for the startup, operation, shutdown, and decommissioning of Soviet-designed nuclear power plants; the safe and secure use of radioactive materials; and the accounting for and protection of nuclear materials. Allocation transfers are legal delegations by one agency of its authority to obligate budget authority and outlay funds to another agency. All financial activity related to these allocation transfers (e.g., budget authority, obligations, outlays) is reported in the financial statements of the parent entity from which the underlying legislative authority, appropriations, and budget apportionments are derived. The NRC receives allocation transfers, as the child, from USAID.

T. Statement of Net Cost

The programs as presented on the Statement of Net Cost are based on the annual performance budget and are described as follows:

Nuclear Reactor and Safety and Security encompasses all NRC efforts to ensure that civilian nuclear power reactor facilities and research and test reactors are licensed and operated in a manner that adequately protects the public health and safety, and the environment, and protects against radiological sabotage and theft or diversion of special nuclear materials. The Nuclear Reactor Safety and Security program contains the following activities: new reactors, reactor licensing tasks, reactor license renewal, international activities, reactor oversight, and incident response.

Nuclear Materials and Waste Safety and Security encompasses all NRC efforts to protect the public health and safety and the environment and ensures the secure use and management of radioactive materials. The Nuclear Materials and Waste Safety and Security program contains the following activities: fuel facilities, nuclear materials users, decommissioning and low-level waste, spent fuel storage and transportation, and high-level waste repository.

For intragovernmental gross costs, the buyers and sellers are both Federal entities. For earned revenues from the public, the buyers of the goods or services are non-Federal entities.

Note 2. FUND BALANCE WITH TREASURY

	2009	2008
Fund Balances		
Appropriated funds	$ 423,724	$ 371,714
Nuclear Waste Fund	24,900	21,764
Other fund types	8	-
Total	$ 448,632	$ 393,478
Status of Fund Balance with Treasury		
Unobligated balance		
Available		
Appropriated funds	$ 74,308	$ 78,877
Unavailable	6,818	113
Obligated balance not yet disbursed	367,498	314,488
Non-budgetary funds with Treasury	8	-
Total	$ 448,632	$ 393,478

The Fund Balance with Treasury consists of unobligated and obligated balance budgetary accounts. It includes Nuclear Waste Fund activity. The Nuclear Waste Fund unobligated balance is $7.6 million and $9.9 million as of September 30, 2009, and 2008, respectively.

Note 3. ACCOUNTS RECEIVABLE

	2009	2008
Intragovernmental		
Fee receivables and reimbursements	$ 4,907	$ 4,692
Receivables with the Public		
Materials and facilities fees-billed	$ 3,316	$ 2,204
Materials and facilities fees-unbilled	122,929	116,162
Other	113	67
Total Receivables with the Public	126,358	118,433
Less: Allowance for uncollectible accounts	(3,141)	(1,749)
Total Receivables with the Public, Net	$ 123,217	$ 116,684
Total Accounts Receivable	$ 131,265	$ 123,125
Less: Allowance for uncollectible accounts	(3,141)	(1,749)
Total Accounts Receivable, Net	$ 128,124	$ 121,376

Note 4. PROPERTY AND EQUIPMENT, NET

Fixed Assets Class	Service Years	Acquisition Value	Accumulated Depreciation and Amortization	2009 Net Book Value	2008 Net Book Value
Equipment	5-8	$ 12,006	$ (10,641)	$ 1,365	$ 1,286
Leased equipment	5-8	1,712	(816)	896	1,239
IT software	5	57,478	(45,522)	11,956	7,181
IT software under development	-	2,227	-	2,227	12,110
Leasehold improvements	20	35,433	(20,706)	14,727	10,081
Leasehold improvements in progress	-	453	-	453	3,578
Total		$ 109,309	$ (77,685)	$ 31,624	$ 35,475

Note 5. OTHER LIABILITIES

	2009	2008
Intragovernmental		
Liability to offset miscellaneous accounts receivable	$ 40	$ 28
Liability for advances from other agencies	88	74
Accrued workers' compensation	1,725	1,710
Accrued unemployment compensation	25	27
Employee benefit contributions	3,611	3,005
Total Intragovernmental Other Liabilities	$ 5,489	$ 4,844
Other Liabilities		
Accrued annual leave	$ 47,271	$ 43,675
Accrued salaries and benefits	23,134	19,683
Contract holdbacks, advances, capital lease liability, and other	7,155	6,929
Grants payable	3,079	661
Total Other Liabilities	$ 80,639	$ 70,948
Total Intragovernmental and Other Liabilities	$ 86,128	$ 75,792

Other liabilities are current except for capital lease liability (Note 7).

Note 6. LIABILITIES NOT COVERED BY BUDGETARY RESOURCES

	2009	2008
Intragovernmental		
FECA paid by DOL	$ 1,725	$ 1,710
Accrued unemployment compensation	25	27
Federal Employee Benefits		
Future FECA	7,628	7,059
Other		
Accrued annual leave	47,271	43,675
Total Liabilities not Covered by Budgetary Resources	56,649	52,471
Total Liabilities Covered by Budgetary Resources	88,107	84,503
Total Liabilities	$ 144,756	$ 136,974

Liabilities Not Covered by Budgetary Resources represents the amount of future funding needed to pay the accrued unfunded expenses as of September 30, 2009, and 2008. These liabilities are not funded from current or prior-year appropriations and assessments, but rather should be funded from future appropriations and assessments. Accordingly, future funding requirements have been recognized for the expenses that will be paid from future appropriations.

Note 7. LEASES

	2009	2008
Assets under capital leases:		
Copiers and booklet maker	$ 1,712	$ 1,712
Accumulated depreciation	(816)	(473)
Net assets under capital leases	$ 896	$ 1,239

Future Lease Payments Due: Fiscal Year	Capital	Operating	2009	2008
2009	$ -	$ -	$ -	$ 32,684
2010	364	32,518	32,882	32,854
2011	284	32,353	32,637	32,637
2012	272	30,236	30,508	30,508
2013	14	22,610	22,624	22,624
2014 and thereafter	-	25,993	25,993	25,993
Total Lease Liability	934	143,710	144,644	177,300
Add: Imputed Interest	60	-	60	107
Total Future Lease Payments	$ 994	$ 143,710	$ 144,704	$ 177,407

The Capital Lease Liability of $934 thousand is included in Other Liabilities (Note 5).

NOTE 8. CUMULATIVE RESULTS OF OPERATIONS

	2009	2008
Liabilities not covered by budgetary resources (Note 6)	$ (56,649)	$ (52,471)
Investment in property and equipment, net (Note 4)	31,624	35,475
Contributions from foreign cooperative research agreements	2,606	3,054
Nuclear Waste Fund	23,703	21,439
Accounts receivable - fees	127,020	120,704
Other	55	34
Cumulative Results of Operations	$ 128,359	$ 128,235

NOTE 9. STATEMENT OF NET COST

For the years ended September 30,	2009	2008
Nuclear Reactor Safety and Security		
Intragovernmental gross costs	$ 238,234	$ 205,183
Less: Intragovernmental earned revenue	(39,307)	(32,710)
Intragovernmental net costs	198,927	172,473
Gross costs with the public	558,664	500,649
Less: Earned revenues from the public	(754,700)	(693,130)
Net costs with the public	(196,036)	(192,481)
Total Net Cost of Nuclear Reactor Safety and Security	$ 2,891	$ (20,008)
Nuclear Materials and Waste Safety and Security		
Intragovernmental gross costs	$ 59,253	$ 54,978
Less: Intragovernmental earned revenue	(6,190)	(6,011)
Intragovernmental net costs	53,063	48,967
Gross costs with the public	186,708	183,241
Less: Earned revenues from the public	(72,270)	(65,729)
Net costs with the public	114,438	117,512
Total Net Cost of Nuclear Materials and Waste Safety and Security	$ 167,501	$ 166,479

NOTE 10. EXCHANGE REVENUES

	2009	2008
Fees for licensing, inspection, and other services	$ 864,155	$ 790,910
Revenue from reimbursable work	8,312	6,670
Total Exchange Revenues	$ 872,467	$ 797,580

Note 11. FINANCING SOURCES OTHER THAN EXCHANGE REVENUE

	2009	2008
Appropriations Used		
Collections were used to reduce the fiscal year's appropriations recognized:		
Funds consumed	$ 993,884	$ 908,330
Less: Collection from fees assessed	(857,839)	(763,640)
Less: Nuclear Waste Funding expense	(46,736)	(46,518)
Total Appropriations Used	$ 89,309	$ 98,172

Funds consumed includes $78.9 million and $72.2 million through
September 30, 2009, and 2008 respectively, of available funds from prior years.

	2009	2008
Non-Exchange Revenue		
Civil penalties	$ 278	$ 1,102
Miscellaneous receipts	108	211
Contra-Revenue	(386)	(1,313)
Total Non-Exchange Revenue	$ -	$ -

	2009	2008
Imputed Financing		
Civil Service Retirement System	$ 11,258	$ 10,239
Federal Employee Health Benefit	19,898	16,589
Federal Employee Group Life Insurance	88	79
Judgements/Awards	963	4
Total Imputed Financing	$ 32,207	$ 26,911

Note 12. TOTAL OBLIGATIONS INCURRED

	2009	2008
Direct Obligations		
Category A	$ 1,022,122	$ 895,751
Exempt from Apportionment	51,660	46,191
Total Direct Obligations	1,073,782	941,942
Reimbursable Obligations	10,297	7,880
Total Obligations Incurred	**$ 1,084,079**	**$ 949,822**

Obligations exempt from apportionment are the result of funds derived from the Nuclear Waste Fund. Category A Obligations consist of NRC appropriations only. Undelivered orders for the Nuclear Waste Fund are $16.1 million and $11.6 million, Salaries and Expenses $276.2 million and $228.4 million, and the Office of the Inspector General $2.3 million and $1.5 million through September 30, 2009, and 2008, respectively.

Note 13. NUCLEAR WASTE FUND

Included in NRC's budget for FY 2009 and 2008 are $49.0 million and $29.0 million, respectively, provided from the Nuclear Waste Fund. Statement of Federal Financial Accounting Standards (SFFAS) No. 27, "Identifying and Reporting Earmarked Funds," lists three defining criteria for an earmarked fund. Generally, an earmarked fund is established by law to use specifically identified financing sources only for designated activities, and the statute provides explicit authority to retain current, unused revenues for future use. Also, the law includes a requirement to account for and report on the receipt and use of the financing sources as distinguished from general revenues.

In 1982, Congress passed the Nuclear Waste Policy Act of 1982 (Public Law 97-425) establishing the Nuclear Waste Fund (NWF) to be administered by the U.S. Department of Energy (DOE) (42 U.S.C. 10222). Given the terms of the statute, the NWF clearly meets the definition of an earmarked fund from DOE's perspective, and DOE does indeed report the NWF as an earmarked fund in its Performance and Accountability Report (PAR).

For the NRC, the NWF transfer is a source of financing; its receipt of NWF funds is a use of NWF resources. The NRC collects no revenue on behalf of the NWF and has no administrative control over it. Furthermore, the Treasury has no separate fund symbol for the NWF under the NRC's agency location code. The receipt and expenditure of NWF money is reported to Treasury under the NRC's primary Salaries and Expenses fund (X0200).

Based on these facts, the NWF is not an earmarked fund from the NRC's perspective. In order to provide additional information to the users of these financial statements, enhanced disclosure of the fund is presented below.

The funding provided to the NRC in FY 2009 and FY 2008 was for the purpose of performing activities associated with DOE's application for a high-level waste repository at Yucca Mountain, NV. These activities included assistance to DOE with the application, review of the application, conduct of thorough safety and security evaluations, preparation of the safety evaluation report, initiation of the inspection program, ensuring that the regulation process was made available to stakeholders and the general public, and providing legal advice and representation for staff reviews and Commission actions.

The NWF amounts received, expended, obligated, and unobligated balances as of September 30, 2009, and 2008 are shown in the following:

	2009	2008
Appropriations received	$ **49,000**	$ 29,025
Expended appropriations	$ **47,062**	$ 48,885
Obligations incurred	$ **51,660**	$ 46,191
Unobligated balances	$ **7,608**	$ 9,853

Note 14. EXPLANATION OF DIFFERENCES BETWEEN THE STATEMENT OF BUDGETARY RESOURCES AND THE BUDGET OF THE U.S. GOVERNMENT

Statement of Federal Financial Standards (SFFAS) No. 7, "Accounting for Revenue and Other Financing Sources," requires the NRC to reconcile the budgetary resources reported on the Statement of Budgetary Resources to the prior fiscal year actual budgetary resources presented in the Budget of the U.S. Government and explain any material differences. The NRC does not have any material differences between the Statement of Budgetary Resources and the Budget of the U.S. Government. The President's Budget with actual results for the NRC has not been published for FY 2009. It is expected to be published in February 2010.

Note 15. RECONCILIATION OF NET COST OF OPERATIONS TO BUDGETARY RESOURCES

For the years ended September 30,	2009	2008
Budgetary Resources Obligated		
Obligations incurred (Note 12)	$ **1,084,079**	$ 949,822
Less: Spending authority from offsetting collections and recoveries	**(40,698)**	(30,578)
Less: Distributed offsetting receipts	**(857,839)**	(763,640)
Net Obligations	**185,542**	155,604
Other Resources		
Imputed financing from costs absorbed by others	**32,207**	26,911
Other	**-**	93,434
Net Other Resources Used to Finance Activities	**32,207**	120,345
Total Resources Used to Finance Activities	**217,749**	275,949
Resources Used to Finance Items not Part of the Net Cost of Operations	**(53,413)**	(19,841)
Total Resources Used to Finance the Net Cost of Operations	**164,336**	256,108
Components of the Net Cost of Operations that will not require or generate resources in the current period	**6,056**	(109,637)
Net Cost Of Operations	$ **170,392**	$ 146,471

Note 16. STATEMENT OF CHANGES IN NET POSITION – ACCOUNTING CHANGE

The NRC is required to recover approximately 90 percent of its budget authority through fee billing and to return the collections to the Treasury. During the first three quarters of FY 2008 when fee revenue was recorded, the NRC also recorded a corresponding liability to the Treasury for the eventual collections. As the actual collections were returned to the Treasury, the liability was reduced. Beginning in the fourth quarter of FY 2008, a change was made to the accounting treatment for recording fee revenue and the corresponding transfer of fee revenue collections to the Treasury. The NRC no longer records the liability to the Treasury when fee revenue is recorded and no longer reduces the liability as the collections are returned to the Treasury. These changes were made to reflect appropriations law and to ensure U.S. Standard General Ledger (USSGL) compliance and consistency. As a result of this change in accounting treatment, in FY 2008 the liability recorded of $93,434 thousand as of FY 2007 was reversed as noted on the FY 2008 Statement of Changes in Net Position.

Required Supplementary Information
Schedule of Budgetary Resources (In Thousands)

For the fiscal year ended September 30, 2009	Salaries and Expenses X0200	Office of Inspector General X0300	Nuclear Facility Fees X5280	Total
Budgetary Resources				
Unobligated balances, brought forward, October 1	$ 78,191	$ 799	$ -	$ 78,990
Recoveries of prior year obligations				
Actual	27,949	422	-	28,371
Budget authority				
Appropriation	1,034,656	10,860	1	1,045,517
Spending authority from offsetting collections				
Reimbursements earned-collected	8,429	-	-	8,429
Reimbursements earned-change in receivables	375	-	-	375
Change in unfilled customer orders-advance received	333	-	-	333
Change in unfilled customer orders-without advance	3,190	-	-	3,190
Subtotal-spending authority from offsetting collections	12,327	-	-	12,327
Total Budgetary Resources	$ 1,153,123	$ 12,081	$ 1	$ 1,165,205
Status of Budgetary Resources				
Obligations incurred (Note 12)				
Direct	$ 1,063,169	$ 10,613	$ -	$ 1,073,782
Reimbursable	10,297	-	-	10,297
Subtotal	1,073,466	10,613	-	1,084,079
Unobligated balance				
Apportioned	65,231	1,468	-	66,699
Exempt from apportionment	7,608	-	1	7,609
Subtotal	72,839	1,468	1	74,308
Unobligated balance, not available	6,818	-	-	6,818
Total Status of Budgetary Resources	$ 1,153,123	$ 12,081	$ 1	$ 1,165,205
Change in Obligated Balance				
Obligated balance, net				
Unpaid obligations, brought forward, October 1	$ 313,573	$ 915	$ -	$ 314,488
Obligations incurred, net	1,073,466	10,613	-	1,084,079
Gross outlays	(989,674)	(9,459)	-	(999,133)
Recoveries of prior year obligations, actual	(27,949)	(422)	-	(28,371)
Change in uncollected customer payments, from Federal Sources	(3,565)	-	-	(3,565)
Obligated balance, net, end of period				
Unpaid obligations	373,554	1,647	-	375,201
Uncollected customer payments, from Federal sources	(7,703)	-	-	(7,703)
Total unpaid obligated balance, net, end of period	$ 365,851	$ 1,647	$ -	$ 367,498
Net outlays				
Gross outlays	$ 989,674	$ 9,459	$ -	$ 999,133
Offsetting collections	(8,762)	-	-	(8,762)
Distributed offsetting receipts	-	-	(857,839)	(857,839)
Net Outlays	$ 980,912	$ 9,459	$ (857,839)	$ 132,532

Auditor's Report

UNITED STATES
NUCLEAR REGULATORY COMMISSION
WASHINGTON, D.C. 20555-0001

OFFICE OF THE
INSPECTOR GENERAL

November 10, 2009

MEMORANDUM TO: Chairman Jaczko

FROM: Hubert T. Bell /**RA**/
                                     Inspector General

SUBJECT: RESULTS OF THE AUDIT OF THE UNITED STATES
                                     NUCLEAR REGULATORY COMMISSION'S FINANCIAL
                                     STATEMENTS FOR FISCAL YEARS 2009 and 2008
                                     (OIG-10-A-05)

The Chief Financial Officers Act of 1990, as amended (CFO Act), requires the Inspector
General (IG) or an independent external auditor, as determined by the IG, to annually
audit the United States Nuclear Regulatory Commission's (NRC) financial statements in
accordance with applicable standards. In compliance with this requirement, Urbach
Kahn & Werlin, LLP (UKW) was retained by the Office of the Inspector General (OIG) to
conduct this annual audit. Transmitted with this memorandum are the following UKW
reports:

- Opinion on the Principal Statements.

- Opinion on Internal Control.

- Compliance with Laws and Regulations.

NRC's Performance and Accountability Report includes comparative financial
statements for FY 2009 and FY 2008.

Objective of a Financial Statement Audit

The objective of a financial statement audit is to determine whether the audited entity's
financial statements are free of material misstatement. An audit includes examining, on
a test basis, evidence supporting the amounts and disclosures in the financial
statements. An audit also includes assessing the accounting principles used and
significant estimates made by management as well as evaluating the overall financial
statement presentation.

UKW's audit and examination were made in accordance with auditing standards generally accepted in the United States of America; *Government Auditing Standards* issued by the Comptroller General of the United States; attestation standards established by the American Institute of Certified Public Accountants; and Office of Management and Budget (OMB) Bulletin No. 07-04, *Audit Requirements for Federal Financial Statements*, as amended. The audit included, among other things, obtaining an understanding of NRC and its operations, including internal control over financial reporting; evaluating the design and operating effectiveness of internal control and assessing risk; and testing relevant internal controls over financial reporting. Because of inherent limitations in any internal control, misstatements due to error or fraud may occur and not be detected. Also, projections of any evaluation of the internal control to future periods are subject to the risk that the internal control may become inadequate because of changes in conditions, or that the degree of compliance with the policies or procedures may deteriorate.

FY 2009 Audit Results

The results are as follows:

Financial Statements

- Unqualified opinion

Internal Controls

- Unqualified opinion

Compliance with Laws and Regulations

- No reportable instances of noncompliance/no substantial noncompliance noted

Office of the Inspector General Oversight of UKW Performance

To fulfill our responsibilities under the CFO Act and related legislation for ensuring the quality of the audit work performed, we monitored UKW's audit of NRC's FY 2009 and FY 2008 financial statements by:

- Reviewing UKW's audit approach and planning.

- Evaluating the qualifications and independence of UKW's auditors.

- Monitoring audit progress at key points.

- Examining the working papers related to planning and performing the audit and assessing NRC's internal controls.

- Reviewing UKW's audit reports to ensure compliance with *Government Auditing Standards* and OMB Bulletin No. 07-04, as amended.

- Coordinating the issuance of the audit reports.

- Performing other procedures deemed necessary.

UKW is responsible for the attached auditors' reports, dated November 6, 2009, and the conclusions expressed therein. OIG is responsible for technical and administrative oversight regarding the firm's performance under the terms of the contract. Our review, as differentiated from an audit in conformance with *Government Auditing Standards*, was not intended to enable us to express, and accordingly we do not express, an opinion on:

- NRC's financial statements.

- The effectiveness of NRC's internal control over financial reporting.

- NRC's compliance with laws and regulations.

However, our monitoring review, as described above, disclosed no instances where UKW did not comply, in all material respects, with applicable auditing standards.

Meeting with the Chief Financial Officer

At the exit conference on November 6, 2009, representatives of the Office of the Chief Financial Officer, OIG, and UKW discussed the results of the audit.

Comments of the Chief Financial Officer

In his response, the Chief Financial Officer (CFO) agreed with the report. The full text of the CFO's response follows this report.

We appreciate NRC staff's cooperation and continued interest in improving financial management within NRC.

Attachment: As stated

cc: Commissioner Klein
       Commissioner Svinicki
       N. Mamish, OEDO
       J. Andersen, OEDO

Independent Auditor's Report on the Financial Statements

INDEPENDENT AUDITOR'S REPORT

Inspector General
United States Nuclear Regulatory Commission

Chairman
United States Nuclear Regulatory Commission

We have audited the accompanying balance sheets of the United States Nuclear Regulatory Commission (NRC), as of September 30, 2009 and 2008, and the related statements of net cost, changes in net position, and budgetary resources (Principal Statements) for the years then ended. We also examined the NRC's internal control over financial reporting as of September 30, 2009 and 2008.

Summary

We concluded that the NRC's fiscal year (FY) 2009 Principal Statements are presented fairly, in all material respects, in conformity with accounting principles generally accepted in the United States of America. We also concluded that the NRC maintained, in all material respects, effective internal control over financial reporting. We noted no reportable instances of noncompliance with laws and regulations and no substantial noncompliance with federal financial management systems requirements, applicable Federal accounting standards, and the United States Government Standard General Ledger (USSGL) at the transaction level.

The following sections (including Appendix A) discuss in more detail: (1) these conclusions and our conclusions relating to other information presented in the Performance and Accountability Report, (2) management's responsibilities, (3) our objectives, scope and methodology, and (4) the current status of prior year findings and recommendations.

Opinion on the Principal Statements

In our opinion, the Principal Statements referred to above present fairly, in all material respects, the financial position of the NRC as of September 30, 2009 and 2008, and its net cost, changes in net position, and budgetary resources for the years then ended, in conformity with accounting principles generally accepted in the United States of America.

Opinion on Internal Control

In our opinion, the NRC maintained, in all material respects, effective control over financial reporting as of September 30, 2009, that provided reasonable assurance that misstatements, losses or noncompliance material in relation to the financial statements

INDEPENDENT AUDITOR'S REPORT, Continued

would be prevented, or detected and corrected, on a timely basis. Our opinion is based on criteria established under 31 U.S.C. 3512 (c), (d), the Federal Managers' Financial Integrity Act (FMFIA).

Compliance with Laws and Regulations

The results of our tests of compliance with laws and regulations disclosed no instances of noncompliance that are required to be reported under *Government Auditing Standards* and OMB Bulletin No. 07-04, *Audit Requirements for Federal Financial Statements*, as amended. Providing an opinion on compliance with laws and regulations was not an objective of our audit and, accordingly, we do not express such an opinion.

Under the Federal Financial Management Improvement Act (FFMIA), we are required to report whether the NRC's financial management systems substantially comply with the federal financial management systems requirements, applicable Federal accounting standards, and the United States Government Standard General Ledger (USSGL) at the transaction level. To meet this requirement, we performed tests of compliance with the provisions of FFMIA section 803(a). The results of our tests disclosed no substantial noncompliance with federal financial management systems requirements, applicable Federal accounting standards, and the USSGL at the transaction level.

Other Information

The information in the Management's Discussion and Analysis section of the NRC's Performance and Accountability Report is not a required part of the Principal Statements, but is supplementary information required by accounting principles generally accepted in the United States of America. We have applied certain limited procedures, which consisted principally of inquiries of management regarding the methods of measurement and presentation of the supplementary information. However, we did not audit the information and express no opinion on it.

The Program Performance and Appendices listed in the Table of Contents are presented for additional analysis and are not a required part of the financial statements. Such information has not been subjected to the auditing procedures applied in the audit of the financial statements and, accordingly, we express no opinion on them.

Management Responsibilities

Management is responsible for (1) preparing the Principal Statements in conformity with accounting principles generally accepted in the United States of America, (2) establishing and maintaining effective internal control over financial reporting, and evaluating its effectiveness, (3) ensuring that the NRC's financial management systems substantially comply with FFMIA, and (4) complying with applicable laws and regulations. NRC management evaluated the effectiveness of NRC's internal control over financial reporting as of September 30, 2009, based on criteria established under FMFIA. NRC management's assurances are included in the Systems, Controls, and Legal Compliance section of the Management's Discussion and Analysis.

INDEPENDENT AUDITOR'S REPORT, Continued

An entity's internal control over financial reporting is a process effected by those charged with governance, management, and other personnel, the objectives of which are to provide reasonable assurance that (1) transactions are properly recorded, processed, and summarized to permit the preparation of financial statements in accordance with U.S. generally accepted accounting principles, and assets are safeguarded against loss from unauthorized acquisition, use, or disposition; and (2) transactions are executed in accordance with the laws governing the use of budget authority and other laws and regulations that could have a direct and material effect on the financial statements.

Objectives, Scope and Methodology

We are responsible for planning and performing our audit to obtain reasonable assurance about whether the financial statements are free of material misstatement. An audit includes examining, on a test basis, evidence supporting the amounts and disclosures in the financial statements. An audit also includes assessing the accounting principles used and significant estimates made by management, as well as evaluating the overall financial statement presentation.

We are responsible for planning and performing our examination to obtain reasonable assurance about whether management maintained, in all material respects, effective internal control over financial reporting as of September 30, 2009. Our examination included obtaining an understanding of NRC and its operations, including internal control over financial reporting; considering NRC's process for evaluating and reporting on internal control over financial reporting which the NRC is required to perform by FMFIA; assessing the risk that a material misstatement exists in the financial statements and the risk that a material weakness exists in internal control over financial reporting; evaluating the design and operating effectiveness of internal control and assessing risk; testing relevant internal controls over financial reporting; and performing such other procedures as we considered necessary in the circumstances. We did not test all internal controls relevant to operating objectives as broadly defined by FMFIA.

Because of inherent limitations in any internal control, misstatements due to error or fraud may occur and not be detected. Also, projections of any evaluation of the internal control to future periods are subject to the risk that the internal control may become inadequate because of changes in conditions, or that the degree of compliance with the policies or procedures may deteriorate.

We are also responsible for testing compliance with selected provisions of laws and regulations that have a direct and material effect on the financial statements. We did not test compliance with all laws and regulations applicable to the NRC. We limited our tests of compliance to those laws and regulations required by OMB audit guidance that we deemed applicable to the financial statements for the fiscal years ended September 30, 2009 and 2008. We caution that noncompliance may occur and not be detected by these tests and that such testing may not be sufficient for other purposes.

We conducted our audit and examinations in accordance with auditing standards generally accepted in the United States of America; *Government Auditing Standards*, issued by the Comptroller General of the United States; attestation standards established by the American Institute of Certified Public Accountants; and OMB Bulletin

INDEPENDENT AUDITOR'S REPORT, Continued

No. 07-04, *Audit Requirements for Federal Financial Statements*, as amended. We believe that our audit and examinations provide a reasonable basis for our opinions.

We noted less significant matters involving the NRC's internal control and its operation, which we have reported to the management of the NRC separately.

Distribution

This report is intended solely for the information and use of the NRC OIG, the management of NRC, OMB, the Government Accountability Office and the Congress of the United States, and is not intended to be and should not be used by anyone other than these specified parties.

Urbach Kahn & Werlin LLP

Arlington, Virginia
November 6, 2009

Appendix A
Status of Prior Year Findings and Recommendations

Our assessment of the current status of the significant deficiency and other reportable condition identified in the prior year audit is presented below:

Prior Recommendation	Type	Fiscal Year 2009 Status
1. The NRC CFO should continue to enhance its procedures for determining accounts payable.	2008 Significant Deficiency	Closed.
2. The NRC CIO should complete its certification and accreditation for the License Fee Billing System.	2008 Substantial Noncompliance with laws and regulations.	Closed.

Management's Response to the Independent Auditor's Report on the Financial Statements

UNITED STATES
NUCLEAR REGULATORY COMMISSION
WASHINGTON, D.C. 20555-0001

OFFICE OF THE
CHIEF FINANCIAL OFFICER

November 7, 2009

MEMORANDUM TO: Stephen D. Dingbaum
                   Assistant Inspector General for Audits
                   Office of the Inspector General

FROM: J. E. Dyer
                   Chief Financial Officer

SUBJECT: AUDIT OF THE FISCAL YEAR 2009 AND 2008 FINANCIAL STATEMENTS

We appreciate the collaborative relationship between the Office of the Inspector General, the auditors and the Office of the Chief Financial Officer in supporting our continuing effort to improve financial reporting. We have reviewed the Independent Auditor's Report of the Agency's Fiscal Year 2009 and 2008 financial statements and are in agreement with it.

cc N. Mamish, AO/OEDO
    J. Arildsen, OEDO
    C. Jaegers, OEDO

Other Accompanying Information

Photo Courtesy of NRC Photo Library

Braidwood Nuclear Power Plant is located in northeast Illinois and is run by Exelon Generation Co., LLC.

Photo Courtesy of NRC Photo Library

NRC Senior Resident Inspector Jim Hickey and Resident Inspector Phil Niebaum in the condensor discharge tunnel at Hatch nuclear plant, near Baxley, in southeastern GA.

Inspector General's Assessment of the Most Serious Management and Performance Challenges Facing the NRC

UNITED STATES
NUCLEAR REGULATORY COMMISSION
WASHINGTON, D.C. 20555-0001

OFFICE OF THE
INSPECTOR GENERAL

September 30, 2009

MEMORANDUM TO: Chairman Jaczko

FROM: Hubert T. Bell /RA/
          Inspector General

SUBJECT: INSPECTOR GENERAL'S ASSESSMENT OF THE MOST
             SERIOUS MANAGEMENT AND PERFORMANCE
             CHALLENGES FACING NRC (OIG-09-A-21)

The *Reports Consolidation Act of 2000* requires the Inspector General of each Federal agency to annually summarize what he or she considers to be the most serious management and performance challenges facing the agency and to assess the agency's progress in addressing those challenges. In compliance with the act, I identified seven management and performance challenges confronting the Nuclear Regulatory Commission that I consider to be the most serious.

Each of the seven challenges identified this year also appeared on my 2008 list. The single difference between the 2008 and 2009 lists is that the new list excludes prior challenge 3, *Implementation of a risk-informed and performance-based regulatory approach*. This challenge was included in my first list of challenges, issued to Congress in January 1998, and remained on the list each year since, with slight variations in wording. I removed the challenge from my list this year because the risk-informed and performance-based regulatory approach is now mature and reflected throughout the agency's regulatory framework.

We appreciate the cooperation extended to us during this evaluation. The agency provided comments on this report, which have been incorporated as appropriate. If you have any questions, please contact Stephen D. Dingbaum, Assistant Inspector General for Audits, at 415-5915 or me at 415-5930.

Attachment: As stated

Electronic Distribution

Edwin M. Hackett, Executive Director, Advisory Committee on Reactor
  Safeguards
E. Roy Hawkens, Chief Administrative Judge, Atomic Safety and
  Licensing Board Panel
Stephen G. Burns, General Counsel
Brooke D. Poole, Jr., Director, Office of Commission Appellate Adjudication
James E. Dyer, Chief Financial Officer
Margaret M. Doane, Director, Office of International Programs
Rebecca L. Schmidt, Director, Office of Congressional Affairs
Eliot B. Brenner, Director, Office of Public Affairs
Annette Vietti-Cook, Secretary of the Commission
R. William Borchardt, Executive Director for Operations
Bruce S. Mallett, Deputy Executive Director for Reactor
  and Preparedness Programs, OEDO
Martin J. Virgilio, Deputy Executive Director for Materials, Waste, Research,
  State, Tribal, and Compliance Programs, OEDO
Darren B. Ash, Deputy Executive Director for Corporate Management
  and Chief Information Officer, OEDO
Vonna L. Ordaz, Assistant for Operations, OEDO
Kathryn O. Greene, Director, Office of Administration
Patrick D. Howard, Director, Computer Security Officer
Cynthia A. Carpenter, Director, Office of Enforcement
Charles L. Miller, Director, Office of Federal and State Materials
  and Environmental Management Programs
Guy P. Caputo, Director, Office of Investigations
Thomas M. Boyce, Director, Office of Information Services
James F. McDermott, Director, Office of Human Resources
Michael R. Johnson, Director, Office of New Reactors
Michael F. Weber, Director, Office of Nuclear Material Safety and Safeguards
Eric J. Leeds, Director, Office of Nuclear Reactor Regulation
Brian W. Sheron, Director, Office of Nuclear Regulatory Research
Corenthis B. Kelley, Director, Office of Small Business and Civil Rights
Roy P. Zimmerman, Director, Office of Nuclear Security and Incident Response
Samuel J. Collins, Regional Administrator, Region I
Luis A. Reyes, Regional Administrator, Region II
Mark A. Satorius, Regional Administrator, Region III
Elmo E. Collins, Jr., Regional Administrator, Region IV

EXECUTIVE SUMMARY

BACKGROUND

The *Reports Consolidation Act of 2000* requires the Inspector General (IG) of each Federal agency to annually summarize what he or she considers to be the most serious management and performance challenges facing the agency and to assess the agency's progress in addressing those challenges.

PURPOSE

In accordance with the act, the IG at the U.S. Nuclear Regulatory Commission (NRC) updated what he considers to be the most serious management and performance challenges facing NRC. The IG evaluated the overall work of the Office of the Inspector General (OIG), the OIG staff's general knowledge of agency operations, and other relevant information to develop and update his list of management and performance challenges. As part of the evaluation, OIG staff sought input from NRC's Chairman, Commissioners, and management to obtain their views on what challenges the agency is facing and what efforts the agency has taken to address previously identified management challenges.

RESULTS IN BRIEF

The IG identified seven challenges that he considers the most serious management and performance challenges facing NRC. The challenges identified represent critical areas or difficult tasks that warrant high-level management attention.

The 2009 list of challenges reflects one change from the 2008 list:

- Prior challenge 3, *Implementation of a risk-informed and performance-based regulatory approach,* was removed as a standalone challenge. This challenge was included in the IG's first list of management challenges, issued to Congress in January 1998,[1] and has remained on the list each year since, with slight variations in wording. Although this regulatory approach – which incorporates risk analysis into regulatory decisions so that NRC and licensee attention can be focused on areas of highest

---

[1] In December 1997, prior to the Reports Consolidation Act of 2000, Congressman Armey requested that Inspectors General independently identify the 10 most serious management problems in their respective agencies to help Congress target key problem areas for attention. NRC's IG complied with the request in January 1998. For subsequent lists, congressional members changed the word "problems" to "challenges." The Reports Consolidation Act of 2000 made this an annual reporting requirement for Federal Inspectors General.

risk – is expected to continue evolving in the years ahead, the approach is now mature and reflected throughout the agency's regulatory framework. Therefore, the IG removed the challenge from the 2009 list and, instead, addresses the issue in narrative about the other challenges, as appropriate.

The chart that follows provides an overview of the seven most serious management and performance challenges as of September 30, 2009.

Most Serious Management and Performance Challenges Facing the Nuclear Regulatory Commission * as of September 30, 2009 (as identified by the Inspector General)	
Challenge 1	Protection of nuclear material used for civilian purposes.
Challenge 2	Managing information to balance security with openness and accountability.
Challenge 3	Ability to modify regulatory processes to meet a changing environment, to include the licensing of new nuclear facilities.
Challenge 4	Oversight of radiological waste.
Challenge 5	Implementation of information technology and information security measures.
Challenge 6	Administration of all aspects of financial management.
Challenge 7	Managing human capital.

The most serious management and performance challenges are not ranked in any order of importance.

CONCLUSION

The seven challenges contained in this report are distinct, yet interdependent relative to the accomplishment of NRC's mission. For example, the challenge of managing human capital affects all other management and performance challenges.

The agency's continued progress in taking actions to address the challenges presented should facilitate achieving the agency's mission and goals.

ABBREVIATIONS AND ACRONYMS

CFR	Code of Federal Regulations
CIP	Construction Inspection Program
COL	combined operating license
CUI	controlled unclassified information
FAIMIS	Financial Accounting and Integrated Management Information System
FOIA	Freedom of Information Act
FY	Fiscal Year
IG	Inspector General
IT	information technology
ITAAC	inspections, tests, analyses, and acceptance criteria
MC&A	material control and accounting
NMMSS	Nuclear Materials Management and Safeguards System
NRC	U.S. Nuclear Regulatory Commission
NSTS	National Source Tracking System
OIG	Office of the Inspector General
ROP	Reactor Oversight Process
T&L	Time and Labor

TABLE OF CONTENTS

I. BACKGROUND

On January 24, 2000, Congress enacted the *Reports Consolidation Act of 2000*, requiring Federal agencies to provide financial and performance management information in a more meaningful and useful format for Congress, the President, and the public. The act requires the IG of each Federal agency to annually summarize what he or she considers to be the most serious management and performance challenges facing the agency and to assess the agency's progress in addressing those challenges.

II. PURPOSE

In accordance with the act's provisions, the NRC IG updated what he considers to be the most serious management and performance challenges facing the agency. The IG evaluated the overall work of OIG, the OIG staff's general knowledge of agency operations, and other relevant information to develop and update his list of management and performance challenges.

In addition, OIG sought input from NRC's Chairman, Commissioners, and management to obtain their views on what challenges the agency is facing and what efforts the agency has taken or planned to address previously identified management and performance challenges.

III. EVALUATION RESULTS

The NRC's mission is to regulate the Nation's civilian use of byproduct, source, and special nuclear materials to ensure adequate protection of public health and safety, promote the common defense and security, and protect the environment. Like other Federal agencies, NRC faces management and performance challenges in carrying out its mission.

Determination of Management and Performance Challenges

Congress left the determination and threshold of what constitutes a most serious management and performance challenge to the discretion of the Inspectors General. As a result, the IG applied the following definition in identifying challenges:

Serious management and performance challenges are mission critical areas or programs that have the _potential_ for a perennial weakness or vulnerability that, without substantial management attention, would seriously impact agency operations or strategic goals.

Based on this definition, in 2009, the IG revised his list of the most serious management and performance challenges facing NRC. The challenges identify critical areas or difficult tasks that warrant high-level management attention. The following chart provides an overview of the seven management challenges. The sections that follow the chart provide more detailed descriptions of the challenges, descriptive examples related to the challenges, and examples of efforts that the agency has taken or are underway to address the challenges.

Most Serious Management and Performance Challenges Facing the Nuclear Regulatory Commission * as of September 30, 2009 (as identified by the Inspector General)	
Challenge 1	_Protection of nuclear material used for civilian purposes._
Challenge 2	_Managing information to balance security with openness and accountability._
Challenge 3	_Ability to modify regulatory processes to meet a changing environment, to include the licensing of new nuclear facilities._
Challenge 4	_Oversight of radiological waste._
Challenge 5	_Implementation of information technology and information security measures._
Challenge 6	_Administration of all aspects of financial management._
Challenge 7	_Managing human capital._

The most serious management and performance challenges are not ranked in any order of importance.

Changes to Management Challenges

The IG identified seven challenges that he considers the most serious management and performance challenges facing NRC. The challenges identify critical areas or difficult tasks that warrant high-level management attention.

This year's list of challenges reflects one change from last year's list:

- Prior challenge 3, *Implementation of a risk-informed and performance-based regulatory approach,* was removed as a standalone challenge. This challenge was included in the IG's first list of management challenges, issued to Congress in January 1998,[2] and has remained on the list each year since, with slight variations in wording. Although this regulatory approach – which incorporates risk analysis into regulatory decisions so that NRC and licensee attention and activities can be focused on areas of highest risk – is expected to continue evolving in the years ahead, the approach is now mature and reflected throughout the agency's regulatory framework. Therefore, the IG removed the challenge from the 2009 list and, instead, addresses the issue in narrative about the other challenges, as appropriate.

---

[2] See footnote 1 for a description of management challenges lists developed in response to congressional requests prior to the Reports Consolidation Act of 2000.

CHALLENGE 1
Protection of nuclear material used for civilian purposes.

NRC is authorized to grant licenses for the possession and use of radioactive materials and establish regulations to govern the possession and use of those materials.

NRC's regulations require that certain material licensees have extensive material control and accounting (MC&A) programs as a condition of their licenses. All other material licensees (including those requesting authorization to possess small quantities of special nuclear materials) must develop and implement plans that demonstrate a commitment to accurately control and account for radioactive materials.

NRC may relinquish its authority to regulate certain radioactive materials and limited quantities of special nuclear material to States. After these States demonstrate that their regulatory programs are adequate to protect public health and safety and are compatible with NRC's program, the States enter into an agreement assuming this regulatory authority from NRC and are called Agreement States.

The issues facing NRC and the agency's actions to address each issue include the following:

Issue: Implement the National Source Tracking System (NSTS), Web-Based Licensing,[3] and the Licensing Verification System[4] to ensure the accurate tracking and control of byproduct material, especially those materials with the greatest potential to impact public health and safety.

> **Action**: NSTS became operational early in 2009. Furthermore, the agency is working to meet its goal of getting Web Based Licensing on line by late 2009 or early 2010. NRC continues to work on the development of the Licensing Verification System.

Issue: Ensure that radioactive material is adequately protected to preclude its use for malicious purposes.

---

[3] A Web-Based Licensing system is intended to serve as a repository for nationwide license information from both NRC and Agreement States.
[4] A License Verification System is planned to interface with NSTS and the Web-Based Licensing system to provide secure, online verification of license and inventory information.

Action: NRC staff recently proposed a final rulemaking to amend its regulations to expand the current NSTS to include Category 3 materials. [5] The staff in its proposal concluded that improving the accountability of certain lower level sources could reduce the possibility of these sources being aggregated to higher activity levels for potential malevolent use. In June 2009, the Commission was unable to reach a majority decision on the staff proposal to expand NSTS to include additional specific licensees that possess sealed sources containing Category 3 threshold quantities of radioactive material. Therefore, the proposed action was not approved.

Issue: Ensure adequate inspections to verify licensees' commitments to their MC&A programs.

Action: NRC is enhancing its inspection programs. Currently, in response to an OIG audit,[6] NRC is working on documenting how risk insights were applied to its reactor MC&A program to determine what types and quantities of materials must be inspected and the frequency of those inspections. Fuel cycle facility MC&A inspections have become a shared responsibility between the Office of Nuclear Material Safety and Safeguards and Region II. The agency is training staff to ensure that there are two MC&A inspectors in headquarters and two in Region II.

Issue: Ensure reliable accounting of special nuclear materials in the NRC and Department of Energy's jointly managed Nuclear Materials Management and Safeguards System (NMMSS).

Action: NRC has been working since 2003 to resolve MC&A issues in response to OIG-03-A15, *Audit of NRC's Regulatory Oversight of Special Nuclear Materials.* On February 7, 2008, NRC approved a final rule that amended its regulations in Title 10, Code of Federal Regulations (10 CFR), Parts 40, 72, 73, 74, and 150, to improve the accuracy of material inventory information maintained in NMMSS. The amendments, effective January 1, 2009, lower the threshold of reportable quantities of special nuclear material and certain source materials to NMMSS, modify the

---

[5] Consistent with guidelines established by the International Atomic Energy Agency, NRC categorizes nuclear materials into five groups based on radioactivity level. Category 1 materials pose the greatest potential for radiological consequences and Category 5 the least.
[6] OIG-03-A-15, *Audit of NRC's Regulatory Oversight of Special Nuclear Materials* (March 23, 2003).

types and timing of submittals to NMMSS, and require licensees to reconcile any material inventory discrepancies that NRC identifies in the NMMSS database. NRC anticipates that these changes to NMMSS reporting requirements will improve the accuracy of material balance (inputs/outputs) information.

<u>Issue</u>: Ensure that Agreement State programs are adequate to protect public health and safety and the environment and are compatible with NRC's program.

<u>Action</u>: NRC conducts about eight reviews per year of Agreement State radiation control programs under NRC's Integrated Materials Performance Evaluation Program using performance indicators to ensure that public health and safety is being adequately protected and that Agreement State programs are compatible with NRC's program. NRC plans to perform a self-assessment of its evaluation program by July 2010.

CHALLENGE 2
Managing information to balance security with openness and accountability.

NRC employees create and work with a significant amount of sensitive information that needs to be protected. Such information includes sensitive unclassified information and classified national security information contained in written documents and various electronic databases.

In addressing continuing terrorist activity worldwide, NRC continually reexamines its information management policies and procedures. NRC faces the challenge of attempting to balance the need to protect sensitive information from inappropriate disclosure with the agency's goal of openness in its regulatory processes. Over the past year, NRC has made various efforts to improve public access to information while protecting sensitive information, including security-related information, from inappropriate disclosure.

The issues facing NRC and the agency's actions to address each issue include the following:

Issue: Be responsive to the numerous information requests from State and local governments, as well as the public, to ensure openness and accountability within the resource constraints of the agency.

> **Action**: The staff have conducted a number of stakeholder outreach efforts to include public meetings on specific regulatory issues and with elected officials regarding issues at facilities within their jurisdiction.

Issue: Manage information in accordance with new Federal Government policies for designating, marking, safeguarding, and disseminating controlled unclassified information (CUI).

> **Action**: In May 2008, the White House issued a Presidential Memorandum conveying requirements for the designation and sharing of controlled unclassified information. The memorandum also established a framework for designating, marking, safeguarding, and disseminating terrorism-related CUI that originates in departments and agencies. Under the CUI framework, all CUI will be categorized into one of three combinations of safeguarding procedures and dissemination

controls, which will be indicated through one of three identified markings: (1) Controlled, Standard; (2) Controlled, Specified; and (3) Controlled Enhanced, Specified.

The CUI framework was created to standardize "Sensitive But Unclassified" practices and thereby improve information sharing. NRC has until 2013 to implement new CUI policies and procedures. Safeguards Information[7] is exempt from the new regulations; therefore, NRC will continue to manage Safeguards Information according to current policies.

Issue: Ensure that sensitive information is handled in accordance with agency policies and procedures for public disclosure.

Action: NRC recognized weaknesses in contracts regarding personal information such as social security numbers and dates of birth, and implemented the use of a contract clause for protecting personal information that may be provided, collected, used, possessed, or processed in the course of performing work under an NRC contract.

Action: NRC implemented new Freedom of Information Act (FOIA) procedures based on guidelines issued in January 2009. In a memorandum to heads of executive departments and agencies, the Attorney General instructed Government workers to apply "a presumption of disclosure" when handling FOIA requests. According to the new guidelines, agencies also need to take affirmative steps to make information public and not wait for specific requests from the public.

Action: In April 2009, NRC issued Management Directive 3.17, *NRC Information Quality Program*, to implement the NRC Information Quality Program, and thereby ensure the quality of information it relies on for decisionmaking or

---

[7] Safeguards Information is a special category of sensitive unclassified information authorized by Section 147 of the Atomic Energy Act of 1954, as amended, to be protected. Although Safeguards Information is sensitive unclassified information, it is protected similar to Government classified confidential information and significantly more than other sensitive unclassified information. Disclosure of Safeguards Information could reasonably be expected to have a significant adverse effect on the health and safety of the public and/or the common defense and security by significantly increasing the likelihood of theft, diversion, or sabotage of materials or facilities subject to NRC jurisdiction.

disseminates to the public. The directive includes guidance on how to make an information correction request and a description of the NRC process for processing Information Correction Requests and appeals.

NRC also revised Management Directive 3.4, *Release of Information to the Public*, to reflect current guidance on the timing of release of documents to the public, add consolidated guidance on the withdrawal of documents from the Agencywide Documents Access and Management System Public Library, and reflect the agency's revised policy on protection and disclosure of information. As part of the revision, NRC separated from the Directive and placed on NRC's internal Web site specific guidance for staff concerning which documents should routinely be released to the public. Placement of this guidance on the Web is intended to allow for frequent updating by all offices and regions.

<u>Issue</u>: Provide external stakeholders with clear and accurate information about regulatory programs and facilitate public participation in the regulatory process.

> <u>Action</u>: The staff solicited public input concerning how NRC could increase public access to security information (e.g., security inspection reports) and held four public meeting around the country to acquire opinions. This information was considered by the NRC staff in preparing SECY-08-0185, *Options for Security Openness, Transparency, and Reactor Oversight Process Improvements*, which was presented to the Commission in late 2008. The Commission reviewed the three options presented in SECY-08-0185, two of which would have made more security information available to the public, and considered the staff's recommendation for one of those two options. The Commission was unable to reach a majority decision on the staff's recommendation; therefore, the proposed action was not approved.

<u>Issue</u>: Review and strengthen programs to protect licensee, vendor, and Government-owned assets (e.g., facility designs, technology descriptions, dual use material and components, classified information) from compromise by foreign sources and industrial espionage and increase awareness of the relationship of these assets to the Nation's economic and industrial base and energy infrastructure.

Action: NRC has recognized the need to ensure technological data involving licensee, vendor, and Government-owned assets is fully protected against potential loss to adversaries. NRC has promulgated orders that provide additional security measures for the protection of these assets.

NRC employees and contractors are required to have a baseline level of security awareness upon entry on duty and the receipt of a security clearance. Others, depending on their job and involvement in the creation and use of protected information, are provided various "role based" training programs, such as classifier's training, training for administrative personnel, declassification training, Secret Internet Protocol Router Network users training, and Sensitive Compartmented Information Access training. The training is layered, targeted, and recurring for those who have specific responsibilities for various types of protected information.

In addition, NRC has increased its information security awareness through the issuance of a variety of agencywide announcements informing staff of the methods employed by those targeting NRC information systems and the corresponding need for employees to heighten their computer security information protection posture.

CHALLENGE 3
Ability to modify regulatory processes to meet a changing environment, to include the licensing of new nuclear facilities.

NRC faces the challenge of maintaining its core regulatory programs while adapting to changes in its regulatory environment. NRC must address a growing interest in licensing and constructing new nuclear power plants to meet the Nation's increasing demands for energy production. As of June 2009, NRC had received 18 Combined Operating License (COL) applications. NRC expects to receive an additional five COL applications through FY 2011.

While responding to the emerging demands associated with licensing and regulating new reactors, NRC must maintain focus and effectively carry out its current regulatory responsibilities, such as inspections of the current fleet of operating nuclear reactors and fuel cycle facilities. NRC intends to increase its safety focus on licensing and oversight activities through risk-informed and performance-based regulation.[8]

The challenges facing NRC and the agency's actions to address each challenge include the following:

New Facilities

Issue: Implement the new Construction Inspection Program (CIP). This includes (1) risk-informing CIP activities to ensure the safe operation of newly constructed nuclear facilities and (2) ensuring that the NRC staff has the necessary knowledge and skill to successfully implement the CIP.

> **Action:** The Office of New Reactors has developed the new CIP in accordance with 10 CFR Part 52. The newly developed inspections, tests, analyses, and acceptance criteria (ITAAC) have been integrated into the Part 52 licensing process to create a design-specific, pre-approved set of performance standards that the licensee must meet and that the Commission must find have been met, before the licensee can load fuel and operate the plant.

---

[8] Risk-informed performance-based regulation incorporates risk analysis into regulatory decisions. This approach is intended to improve the regulatory process by focusing both NRC licensee attention and activities on the areas of highest risk.

Additionally, the agency has issued and revised a number of Inspection Manual Chapters and procedures to implement the new ITAAC process.

NRC has revised Inspection Manual Chapter 1252, *Construction Inspector Training and Qualification Program,* to ensure that the agency is effectively preparing inspectors to implement the new CIP. The agency will monitor the effectiveness of the training program as inspections of the new construction projects begin.

Issue: As the public's demand for new energy sources continues, NRC must ensure that the process for reviewing applications for new facilities focuses on safety and effectiveness.

> **Action:** NRC's preparations have been focused on issuing reactor design certifications, revising the regulation that governs early site permits, and engaging in ongoing interactions with nuclear plant designers and utilities regarding prospective new reactor applications and licensing activities. In April 2009, the Office of New Reactors developed a set of goals with the purpose of enhancing the agency's ability to plan and implement its reviews more effectively in a dynamic environment resulting from changes in the applicants' business strategies.

> NRC is taking a "design-centered review approach" to optimize the COL application review process. Part of the license review process includes conducting performance-based vendor inspections and quality assurance/quality control audits.

Existing Fleet

Issue: NRC's license renewal and power uprate review processes must effectively focus on an applicant's ability to ensure the continued safe operation of the plant.

> **Action:** For planning purposes, NRC continues to work with plant licensees to develop a schedule of anticipated requests for license renewals and power uprates. The agency has also implemented a number of recommendations to improve the license renewal review and power uprate processes to include closer management oversight. For license renewal reviews, the agency has updated report-writing guidance to include management expectations and report-writing

standards. For power uprate reviews, the agency has developed a training module for technical reviewers and project managers that is specifically focused on writing or contributing to a safety evaluation.

<u>Issue</u>: Respond to a heightened public focus on license renewals resulting in contested hearings.

> <u>Action</u>: NRC has open dialogs with the industry, licensees, and stakeholders, and appropriate comments have been incorporated into new inspection procedures. Additionally, the license renewal process allows stakeholders to request a hearing in order to present their concerns.

<u>Issue</u>: Ensure the ability to identify emerging operating and safety issues at all plants including issues associated with license renewal and power uprate; consistently apply regulatory and review changes in response to these emerging issues across the existing fleet of reactors.

> <u>Action</u>: NRC continues to make changes to its regulatory programs based on emerging operational and safety issues related to license renewal and power uprate. For example, as a result of identified weaknesses in the power uprate program, Inspection Procedure 71004 was revised to provide additional guidance on inspection planning, implementation, and documentation. Annually, agency staff communicate the status of the license renewal and power uprate programs to the Commission.

<u>Issue</u>: Establish and maintain effective, stable, and predictable regulatory programs or policies for all programs.

> <u>Action</u>: NRC continues to interface with stakeholders, develop regulatory policy, update rules and technical guidance, provide technical leadership and management for the Reactor Oversight Process (ROP), and support the development of programmatic changes when needed. Additionally, the ROP features an annual assessment process which is used to revise the process as necessary.

CHALLENGE 4
Oversight of radiological waste.

NRC regulates spent nuclear fuel generated from commercial nuclear power reactors, which is referred to as high-level radioactive waste. NRC faces significant issues involving the potential licensing of the proposed Yucca Mountain repository for storing high-level radioactive waste. Additional challenges in the high-level waste area include the interim storage of spent nuclear fuel, certification of storage and transportation casks, and the oversight of decommissioned reactors and other nuclear sites.

Additionally, the amount of low-level waste continues to grow; however, no new disposal facilities have been built since the 1980s and unresolved issues will multiply as the once-operational disposal facilities shut down.

The challenges facing NRC and the agency's actions to address each challenge include the following:

Issue: Address increasing quantities of radiological waste requiring interim or permanent disposal sites.

> **Action:** NRC developed and implemented a risk-informed decisionmaking framework in connection with a wide range of nuclear waste storage issues. The NRC has conducted reviews using the framework for dry cask storage systems and concluded that such systems provide a safe means to store spent nuclear fuel with exceedingly low risk. NRC has met with Agreement States and industry to discuss guidance on interim storage of Class B/C waste.[9] Stakeholder outreach is an integral part of NRC's low level waste strategic assessment.

Issue: Address issues regarding the license application to construct a high-level radioactive waste repository at Yucca Mountain.

> **Action:** The NRC is continuing to review the Yucca Mountain license application submitted by the Department of Energy in June 2008, and is conducting high-level waste licensing activities to ensure public health and safety and protection of the environment. Due to the unprecedented

---

[9] Classes A, B, and C radioactive waste present increasing levels of risk of disposal, with Class C waste posing the greatest risk.

number of contentions filed, the Department of Energy's decision not to submit the required Environmental Impact Statement, and the staff's limited resources, the agency has stated that the review proceedings will not meet the 10 CFR Part 2, Appendix D schedule. In 2009, NRC issued the final requirements in 10 CFR Part 63 to align agency regulations to new Environmental Protection Agency standards for radiation protection at a high-level waste repository.

Issue: Oversight of low-level waste disposal, including low-level radioactive waste disposal sites.

Action: NRC modified Inspection Procedure 84900 to address long-term storage of Class B and C waste. As recommended in the *Low-Level Waste Strategic Assessment*, NRC issued Regulatory Issue Summary 2008-32, which consolidated previous NRC low-level waste guidance and communicated to licensees that the NRC's staff position continues to be that low-level waste storage must meet NRC requirements and that when constructing new low-level waste storage facilities the regulations for evaluating proposed changes to facilities must be met.

Issue: Oversight of nuclear waste issues associated with the decommissioning and cleanup of nuclear reactor sites and other facilities.

Action: NRC continues to hold public meetings with stakeholders and licensees to explore safe and secure storage options associated with decommissioning of plants, such as transitioning from spent pool storage to dry cask storage. NRC continues to oversee the 14 power reactors currently undergoing decommissioning. NRC staff published NUREG-1307, *Report on Waste Burial Charges*, which provides updated low-level waste disposal costs for reference pressurized water reactor and boiling water reactor based on estimated disposal volumes.

> ## CHALLENGE 5
> ### *Implementation of information technology and information security measures.*

NRC needs to continue upgrading and modernizing its information technology (IT) and security capabilities both for employees and for public access to the regulatory process. Recognizing the need to modernize, the Office of Information Services established goals to improve the productivity, efficiency, and effectiveness of agency programs and operations, and enhance the use of information for all users inside and outside the agency. NRC also needs to ensure that system security controls are in place to protect the agency's information systems against misuse.

The issues related to this challenge and the agency's actions to address each issue include the following:

<u>Issue</u>: Upgrade and manage IT activities to improve the productivity, efficiency, and effectiveness of agency programs and operations.

> <u>Action</u>: A specialized team reporting to the Director of the Office of Information Services Infrastructure and Computer Operations Division was established. This team is responsible for planning and coordination activities to ensure the agency's IT infrastructure is sufficient to support growth and program needs.

> <u>Action</u>: In the second quarter of FY 2009, the Office of Information Services hosted a 1-day IT summit for NRC offices and regions to engage in discussions with offices and regions to provide a better understanding of the agency's IT program, raise an appreciation for the necessary and crucial planning needed to properly execute technology projects, whether they are led from a program office or the Office of Information Services, and to verify the current plans and directions for technology modernization across the agency. A variety of presentations enhanced understanding and encouraged discussion of IT project interrelationships and explained the need for a clear, agreed-upon strategy and path forward on all agency IT projects.

> <u>Action</u>: An aggressive implementation schedule was developed to upgrade the existing IT environment and to bring new technologies to NRC. Some of the projects under

development include strategies, methods, and tools used to capture, manage, store, preserve, and deliver content and documents related to organizational processes.

Issue: Provide laptop computers with enhanced functionality, security, and support.

> **Action**: An agency laptop standard and security policy was developed and published on the NRC intranet and distributed via an announcement. The Customer Support Center offers encrypted thumb drives as well as a laptop loaner program. Standard policies for the use of commercial wireless devices, services, and technologies have also been implemented. The IT infrastructure was expanded to support 1,000 Blackberry devices.

Issue: Ensure that information systems and assets are protected.

> **Action**: The Computer Security Office has taken action on identified vulnerabilities. Such actions include (1) certifying and accrediting 89 percent of the agency's systems that are reported to the Office of Management and Budget under the agency's Federal Information Security Management System, (2) initiating a continuous monitoring system to evaluate IT security controls of agency IT systems to provide assurance that systems remain secure after having been authorized to operate, and (3) publishing IT security policy to address current agency needs to include encryption of data at rest, encryption of data in transmission, and use of thumb drives.

> **Action**: The agency is providing a secure network for authorized users to access safeguards information documents electronically. This system will cut down on the need to print documents and will enable the management and collaboration of safeguards documents in a centralized electronic document management system.

Issue: Ensure that plans for a cyber security inspection program are developed and implemented.

> **Action**: The staff plans to develop an inspection procedure for conducting cyber security inspections at nuclear power plants and hold training for NRC cyber security inspectors. The inspections are planned to be conducted between calendar year 2012 and 2016.

> CHALLENGE 6
> *Administration of all aspects of financial management.*

NRC management is responsible for establishing and maintaining effective internal controls and financial management systems that meet the objectives of several statutes including the Federal Managers' Financial Integrity Act. This act mandates that NRC establish controls that reasonably ensure that (1) obligations and costs comply with applicable law; (2) assets are safeguarded against waste, loss, unauthorized use, or misappropriation; and (3) revenues and expenditures are properly recorded and accounted for. This act encompasses program operational, and administrative areas, as well as accounting and financial management.

In addition, NRC's management of its expanded grant program must be conducted in accordance with Federal regulations, which includes ensuring that funds are distributed and used as intended.

The issues related to this challenge and the agency's actions to address each issue include the following:

Issue: Replace the agency's current financial systems, which are obsolete, overly complex, and inefficient.

> **Action**: On May 8, 2009, the Chief Financial Officer announced that the agency had selected Momentum Financials as the software for the Financial Accounting and Integrated Management Information System (FAIMIS) implementation project. In June 2009, the agency began the configuration and integration phase of the project. The planned "go-live" date is October 1, 2010. To ensure that FAIMIS performs as intended, the agency will conduct final user acceptance testing and end-to-end testing in parallel with the NRC legacy systems during the 6 months prior to the October "go-live" date.

> **Action**: On July 20, 2009, NRC implemented e-Travel, a governmentwide initiative to improve travel operations and management. The paperless system automates travel documentation and approval, funds certification, and booking of travel reservations. Currently, e-Travel includes local and temporary duty travel. The agency plans to implement specialized travel, such as foreign and premium class travel, by December 2010.

Action: NRC plans to implement an upgrade to the Time and Labor System (T&L) during the second quarter of FY 2010. The upgrade will provide a modern, Web-enabled version of the existing PeopleSoft T&L software. The system will also provide the capability for electronic workflow and approval of employee timesheets.

Issue: Respond to Commission direction and implement recommendations of the Advisory Group on Budget Formulation and Financial Plan Reporting (Advisory Group). This issue encompasses both budget formulation and budget execution.

Action: In response to direction from the Commission to improve the agency's budget formulation process, the agency has undertaken efforts to implement a more top-down, programmatic-based budget process. Improvements include an update of the budget structure, which is being implemented in a two-phased approach. The agency is identifying product lines and specific products for each product line during formulation of the FY 2011 and 2012 budgets, respectively. The ultimate goal of the budget structure is to integrate budget formulation, execution, and performance information to support the assessment of the efficiency and effectiveness of agency programs, products, and activities.

Action: Based on the Advisory Group's reviews and initial recommendations, the Office of the Chief Financial Officer (OCFO) and the Office of the Executive Director for Operations (OEDO) made a number of improvements to the FY 2009 budget execution process. These improvements include the CFO/EDO periodic budget briefings, an earlier mid-year review exercise, increased focus on identifying and processing deobligations, quarterly Advanced Procurement Plan updates, and the use of IT to enhance communication and efficiency. A recent Advisory Group report and Commission direction included recommendations for additional improvements to the budget execution process. In general, the level of process maturity for budget execution lags behind that of budget formulation by a few years. Additional challenges in this area may result from possible future Continuing Resolutions, which can impact the timing of receipt of future appropriations.

Issue: Manage the agency's expanded grant program to ensure funds are efficiently and effectively distributed and used as intended.

Action: In response to legislative and congressional direction over the past several years, NRC has initiated and implemented a new education grant program. In FY 2007, NRC was authorized to distribute $5M in education grants, and for both FY 2008 and 2009, this figure increased to $20 million. Prior to FY 2007, NRC's financial assistance program was much smaller; for example, NRC provided about $1.5 million in assistance during FY 2006, and $564,000 in FY 2005. To implement its expanded grant program, NRC has hired several grant experts from other Federal agencies and is working to establish and document a process for announcing grants, reviewing applications, and administering these types of grants. NRC also is conducting a Lean Six Sigma[10] review of the agency's process for awarding grants to reduce the overall time for processing grants. Another Lean Six Sigma goal is to implement a formal, electronic tracking and reporting system that minimizes data inconsistencies.

In addition to the issues noted above, the agency has taken several steps to meet the challenge of administering all aspects of financial management. Those steps include completing the requirements to receive a certification and accreditation over the License Fee Billing System, instituting a new process to estimate the accounts payable balance, evaluating the expansion of the cross-servicing effort to other NRC financial activities, and streamlining the financial reports preparation process for account reconciliation and financial statement generation.

---

[10] Lean Six Sigma is a methodology for improving business processes that combines the strategies and tools from two other business process improvement methodologies focused on (1) reducing process time and resources by eliminating unnecessary delays and steps, and (2) identifying and reducing specific sources of process variation. Combining these two complimentary methodologies, in conjunction with a successful implementation, is thought to result in a process that is faster, takes less resources, is more consistent, and, therefore, more predictable.

CHALLENGE 7
Managing human capital.

NRC's human capital needs are changing due to the receipt of applications to construct and operate the next generation of nuclear reactors and to increase the number of fuel cycle facilities. To effectively manage human capital as these changes progress, while continuing to accomplish the agency's mission, NRC must continue to implement the following initiatives:

- Timely personnel security adjudication.
- Space planning.
- Recruitment, training, and knowledge management.
- Optimal use of resources.

The issues related to this challenge and the agency's actions to address each issue include the following:

Issue: Timely personnel security adjudication. Work start dates for NRC employees, contractors, and licensees are frequently delayed for months at a time due to the time-consuming personnel security adjudication process currently in place for granting access authorization.

> **Action**: The agency is reviewing its hiring process for external applicants, which includes the entire hiring and security process that occurs from identification of an active vacancy through the entrance on duty date, and plans to develop recommendations to expedite the process during FY 2010.

> **Action**: In accordance with Executive Order 13467, *Reforming Processes Related to Suitability for Government Employment, Fitness for Contractor Employees, and Eligibility for Access to Classified National Security Information* dated June 30, 2008, NRC has developed reciprocity processes and procedures to accept applicable investigations and adjudications conducted by other Federal agencies.

> **Action**: The HR Recruitment Activity Tracking System was modified to include security processing and adjudication status information. Reports from this system are shared with the program offices to keep managers informed of the status of their new hires.

Issue: Space planning. NRC must continue to accomplish the agency's mission and communicate effectively with staff located in multiple office locations.

> **Action**: NRC is implementing a Headquarters Strategic Housing Plan designed to meet space needs through FY 2013. The agency expects to begin consolidating staff by occupying a new permanent building in close proximity to the White Flint Complex in FY 2012. Furthermore, most NRC regional offices are seeking new office space to accommodate additional staff in order to meet increased workload demands.

> **Action**: To ensure the agency maintains its sense of one community while employees are in multiple locations, a Staying Connected Working Group that includes representatives from various NRC Offices has been meeting regularly since 2008. Its role is to confirm that employees in interim buildings are receiving the services they need to perform their work and maintain workplace satisfaction. Among other accomplishments, the working group has facilitated onsite demonstrations of virtual meeting service for employees in interim buildings and arranged to increase shuttle service among headquarters buildings. The group also is looking at the uses of videoconferencing and other electronic methods to facilitate connectivity, and it is developing a "Staying Connected" Web page.

> **Action**: On January 1, 2009, Region III acquired an additional 11,028 square feet of leased office space to consolidate offices and improve the readiness and capabilities by further enhancing the integration of safety and security with key emergency and communications systems.

Issue: Recruitment, training, and knowledge management. NRC must continue to address anticipated increased workload demands and retirements.

> **Action**: NRC is refining the agency's human capital program through the following initiatives: (1) streamlining and enhancing the hiring process through use of emerging technologies such as the centralized hiring tools available through the USAJobs.Gov Web site and implementation of Lean Six Sigma process change recommendations; (2) refining work life programs intended to improve employee satisfaction and wellness; (3) implementing a Leaders

Academy that provides contemporary management and leadership training and continued development to current, future and potential NRC leaders; and (4) using advanced training methods to improve skills, target individual learning styles, reduce travel, and reduce time to competency.

Action: NRC is implementing knowledge management strategies[11] that include mentoring, early replacement hiring, and rehiring annuitants with or without use of a pension offset as applicable.[12] The agency also has developed a knowledge management Web site, expressly for the purpose of retaining knowledge before key employees are promoted or retire.

Action: In response to legislative and congressional direction, NRC recently implemented a $20-million per year education grant program to support and develop the educational infrastructure necessary to allow the Nation to safely move its nuclear energy initiatives forward. Funds are used to support courses and curricula relevant to careers in the nuclear field and to provide scholarships, fellowships, and faculty development to benefit the nuclear sector. While these grants are expected ultimately to benefit the entire nuclear profession by increasing the pool of competent, qualified workers to work in the nuclear field, NRC also benefits from an increased applicant pool from which to draw.

Issue: Optimize utilization of resources to address the change in agency workload resulting from various states becoming Agreement States.

Action: Regional and program offices are working cooperatively within the budget process to assure resources are allocated to address changing workloads. The agency has also been evaluating the impacts on projected workload over the next several years.

---

[11] Knowledge management involves capturing critical information and making the right information available to the right people at the right time to assure that knowledge and experience of the current staff is passed on to the next generation of NRC staff.

[12] This flexibility allows NRC to rehire a retiree to fill a position at full pay if the agency has experienced difficulty in filling a position, or if a temporary emergency exists.

IV. CONCLUSION

The seven challenges contained in this report are distinct, yet are interdependent to accomplishing NRC's mission. For example, the challenge of managing human capital affects all other management and performance challenges.

The agency's continued progress in taking actions to address the challenges presented should facilitate achieving the agency's mission and goals.

SCOPE AND METHODOLOGY

This evaluation focused on the IG's annual assessment of the most serious management and performance challenges facing the NRC. The challenges represent critical areas or difficult tasks that warrant high level management attention. To accomplish this work, the OIG focused on determining (1) current challenges, (2) the agency's efforts to address the challenges during FY 2009, and (3) future agency efforts to address the challenges.

OIG reviewed and analyzed pertinent laws and authoritative guidance, agency documents, and OIG reports, and sought input from NRC officials concerning agency accomplishments relative to the challenge areas and suggestions they had for updating the challenges. Specifically, because challenges affect mission critical areas or programs that have the potential to impact agency operations or strategic goals, NRC Commission members, offices that report to the Commission, the Executive Director for Operations, and the Chief Financial Officer were afforded the opportunity to share any information and insights on this subject.

OIG conducted this evaluation from June through August 2009. The major contributors to this report were Anthony Lipuma, Deputy Assistant Inspector General for Audits; Steven Zane, Team Leader; Beth Serepca, Team Leader; Sherri Miotla, Team Leader; and Judy Gordon, Quality Assurance Manager.

Reactor installation containment area.

Photo Courtesy of NRC Photo Library

Management Decisions and Final Actions on OIG Audit Recommendations

Oyster Creek Nuclear Generating Station is located south of Toms River, NJ. It is run by Exelon Generation Co., LLC.

Photo Courtesy of NRC Photo Library

Chairman Jazcko (then Commissioner) visiting Donald C. Cook Nuclear Plant in Berrion County, MI in April 2009.

Management Decisions Not Implemented Within 1 Year

For the OIG audit reports listed in the following tables, the NRC made management decisions before October 1, 2008. As of September 30, 2009, NRC had not taken final action, including OIG final review and closure, on some issues. Completion of the activities listed in the column "Actions Pending" will complete agency action on the listed OIG audit and evaluation recommendations.

Government Performance and Results Act: Review of the Fiscal Year 1999 Performance Report (OIG-01-A-03)

February 23, 2001

The Office of the Inspector General (OIG) of the U.S. Nuclear Regulatory Commission (NRC) conducted this audit at the request of the chairman of the Senate Committee on Governmental Affairs to determine whether NRC's fiscal year (FY) 1999 performance data were valid and reliable and whether the FY 2000 performance data would be more valid and reliable. The audit found that, while the NRC was improving and strengthening its performance reporting process, as interim policy guidance, the agency needed to institute management control procedures to produce valid and reliable data. The agency should then institutionalize the procedures in an NRC management directive (MD).

Open Recommendations	Actions Pending
1. Develop an NRC management directive (MD) to provide the management controls needed to ensure that the NRC produces credible Government Performance and Results Act (GPRA) documents.	The NRC issued interim guidance for performance management and reporting performance information in July 2001, consistent with GPRA requirements. In July 2002, the NRC issued a new MD 4.8 and associated Handbook, titled, "Performance Measurements," for intraagency review and comment. Staff subsequently decided that the agency should address performance measurement in the broader context of budget and performance integration. Therefore, the NRC decided to incorporate MD 4.8 into a revision of MD 4.7 and Handbook, which will be titled, "Planning, Budgeting, and Performance Management." The revised MD 4.7 and Handbook will clarify the roles and responsibilities in setting the agency's strategic direction, determining planned activities and resources, measuring and monitoring performance, and assessing performance.
3.. Include guidance on reporting unmet goals in both the management directive and the interim policy guidance on implementing GPRA initiatives.	
	In August 2007, the Commission directed the Chief Financial Officer, in coordination with staff, to provide options for improving the agency's budget formulation process. The staff developed and implemented a new top-down budget process in formulating the agency's FY 2010 and FY 2011 budgets. Subsequently, the staff considered lessons learned from the NRC task force that reviewed the agency's budget formulation process.
	As a result of lessons learned from the budget process, the agency decided that it would address its strategic planning process and performance assessment in separate management directives. MD 4.7 and Handbook will address roles and responsibilities in the agency's budget formulation process.
	Based on the task force's current schedule for issuing guidance on the agency's budget formulation process, the staff expects to publish MD 4.7 and Handbook in January 2010.

Review of the NRC's Handling and Marking of Sensitive Unclassified Information (OIG-03-A-01)

October 16, 2002

The OIG conducted this audit to assess NRC's program for handling, marking, and protecting of Official Use Only (OUO) information, a category of sensitive unclassified information. The audit found that NRC's program and guidance for the handling and marking of sensitive unclassified information may not adequately protect OUO information from inadvertent public disclosure. The audit also found that the agency does provide training on a regular basis to all NRC employees and contractors on handling and protecting sensitive unclassified information.

Open Recommendations	Actions Pending
1. Update the guidance for OUO documents to require clear identification of sensitive unclassified information to prevent its inadvertent disclosure. 2. Mandate consistent use of defined markings on documents containing OUO information and clarify the markings that should be used on sensitive unclassified information.	Agency corrective actions require issuance of a revised MD covering sensitive unclassified, nonsafeguards information (SUNSI) and a new MD covering safeguards information (SGI). The NRC issued MD 12.7, "NRC Safeguards Information Security Program," on SGI on June 25, 2009. The revision of SUNSI is on hold pending the issuance of standard Federal guidance on Controlled Unclassified Information (CUI) by the National Archives and Records Administration (NARA), which is the executive agent for implementing the CUI policy. The NRC will revise SUNSI policy to align it with the CUI guidance.

Audit of the NRC's Regulatory Oversight of Special Nuclear Materials (OIG-03-A-15)

May 23, 2003

OIG conducted this audit to determine whether the NRC adequately ensures that its licensees control and account for special nuclear material (SNM). The audit found that NRC's current level of oversight of licensees' material control and accounting (MC&A) activities does not provide adequate assurance that all licensees properly control and account for SNM. The audit reported that the NRC performs only limited inspections of licensees' MC&A activities and thus cannot ensure the reliability of data in the Nuclear Materials Management & Safeguards System (NMMSS). The U.S. Department of Energy manages this computer database and shares it with the NRC as the national system for tracking certain private- and Government-owned nuclear materials.

Open Recommendations	Actions Pending
1. Conduct periodic inspections to verify that material licensees comply with MC&A requirements, including, but not limited to, visual inspections of licensees' SNM inventories and validation of report information.	The NRC issued a revised rule, effective January 1, 2009, to address the issue of NMMSS records, and is working on a separate revision of MCEA regulations for fuel cycle facilities. The final rule will consolidate MC&A regulations and the licensing process. The technical basis for the rulemaking will include the basis for making the MC&A program risk informed and explain how the rulemaking will be applied to the program.
3. Document the basis of the approach used to risk inform NRC's oversight of MC&A activities for all types of materials licensees.	As part of the revised Fuel Cycle Facility Oversight Process (FCOP) currently under development, the NRC is revising inspection plans applicable to fuel facilities, including MC&A inspections. The FCOP is designed to be a more risk-informed and performance-based process. As part of this process, the NRC may revisit inspection resources and frequencies for all types of materials licensees' MC&A inspections for SNM.

Audit of the NRC's Incident Response Program (OIG-04-A-20)

September 16, 2004

OIG conducted this audit to determine whether the NRC performs its incident response (IR) program in a timely and effective manner, provides adequate support to licensees, and maintains readiness and qualifications of staff. The audit found that, while the NRC has improved its program since the Three Mile Island 2 accident on March 29, 1979, the agency needs to do more to ensure that the program is performed consistently, is more fully understood by licensees, and maintains a well-defined process for demonstrating that staff are qualified and ready to respond.

Open Recommendations	Actions Pending
4. Periodically review regional incident response programs to ensure NRC's incident response program is carried out consistently across the agency.	Recommendation 4 will be closed when the NRC completes the initial regional review. As of August 2009, the staff has completed the IR program assessments at Regions I, II, IV, and headquarters. The agency published the assessment results in a report to the Office of Nuclear Security and Incident Response's Deputy Director for IR. The report can be found in the Agencywide Documents Access and Management System (ADAMS) under Accession Nos. ML082100586 for Region I, ML073240937 for Region II, ML083390799 for Region IV, and ML092230718 for headquarters. The assessment for Region III is currently scheduled to coincide with Region III's participation in the Braidwood Exercise on March 22, 2010.

Independent Evaluation of the NRC's Implementation of the Federal Information Security Management Act for Fy 2004 (OIG-04-A-22)

September 30, 2004

This was an independent evaluation of the NRC's implementation of the Federal Information Security Management Act (FISMA) for FY 2004. The review found that, while the NRC had made improvements to its automated information security program, the agency still needs to make additional improvements.

Open Recommendations	Actions Pending
5. Recertify and reaccredit the NRC Data Center/Telecommunications System (DC/T).	The authority to operate (ATO) for the telecommunications system was granted on August 21, 2009 (see ADAMS Accession No. ML091270317). There are a total of 29 plans of action and milestones (POA&Ms) for the system, and 11 are currently open. The agency is currently scheduling activities and will report completion dates to the Office of Management and Budget (OMB) as part of the FISMA reporting requirements period which ends December 31, 2009.

System Evaluation of the Integrated Personnel Security System (OIG-05-A-08)

January 14, 2005

OIG conducted this evaluation as part of its review of the NRC's implementation of FISMA for FY 2004. The objective was to review and evaluate the management, operational, and technical controls for the integrated personnel security system (IPSS), which replaced NRC employee security information contained in paper files and in a less-capable automated data system. The review found that the IPSS security test and evaluation were not comprehensive and independent, security documentation was not always consistent with National Institute of Standards and Technology guidelines, and security protection requirements were not consistent within the security documentation.

Open Recommendations	Actions Pending
1. Recertify and reaccredit IPSS based on an independent, comprehensive, and fully documented assessment of all management, operational, and technical controls.	The ATO was granted on September 18, 2009.
2. Update the IPSS risk assessment report to include listed changes.	The IPSS risk assessment was updated. The ATO was granted on September 18, 2009.
3. Update the IPSS system security plan to include listed changes.	The IPSS system security plan was updated. The ATO was granted on September 18, 2009.
4. Update the IPSS system security plan to include a section on planning for security in the life cycle and a section on incident response capability.	The IPSS system security plan was updated. The ATO was granted on September 18, 2009.
5. Update the IPSS system security plan to describe all controls currently in place. In-place controls are those marked at least at Level 3 in the self-assessment and that were documented as passed in the last security test and evaluation report (or in any test and evaluation on controls added since publication of that report).	The IPSS system security plan was updated. The ATO was granted on September 18, 2009.
8. Update the IPSS system security plan and IPSS self-assessment to consistently define the protection requirements (confidentiality, integrity, and availability).	The IPSS system security plan was updated. The ATO was granted on September 18, 2009.

Audit of the NRC's Budget Formulation Process (OIG-05-A-09)

January 31, 2005

OIG conducted the audit to determine whether the budget formulation portion of the NRC's planning, budgeting, and performance management process is effectively used to develop and collect data to align resources with strategic goals and is efficiently and effectively coordinated with program and support offices. The audit found that the NRC effectively develops and collects data to align resources with strategic goals, prepares the budget in alignment with the Strategic Plan, and successfully conducts OMB-required program assessment rating tool evaluations. The audit also found that the agency needed additional internal coordination and communication efforts.

Open Recommendations	Actions Pending
1. Clarify the roles and responsibilities of the Chief Financial Officer and the Executive Director for Operations in the budget formulation process.	A revision of MD 4.7 and Handbook, "Planning, Budgeting, and Performance Management," will clarify roles and responsibilities and document the budget formulation process, including decisionmaking, and will provide for a logical, comprehensive sequencing of events for obtaining early Commission direction and approval.
2. Document the decisionmaking process and the roles and responsibilities of the program review committee.	In August 2007, the Commission directed the Chief Financial Officer, in coordination with staff, to provide options for improving the agency's budget formulation process. The staff developed and implemented a new top-down budget process in formulating the agency's FY 2010 and FY 2011 budgets. Subsequently, the staff considered lessons learned from the NRC task force that reviewed the agency's budget formulation process.
3. Document the budget formulation process to ensure a logical, comprehensive sequencing of events that provides for obtaining early Commission direction and approval.	As a result of lessons learned from the budget process, the agency decided to address its strategic planning process and performance assessment in separate management directives. MD 4.7 and Handbook will address roles and responsibilities in the agency's budget formulation process.
	Based on the task force's current schedule for issuing guidance on the agency's budget formulation process, the agency expects to publish MD 4.7 and Handbook in January 2010. (MD 4.7 and Handbook will also address the decisionmaking roles and responsibilities of the program review committee.)

Audit of the NRC's Telecommunications Program (OIG-05-A-13)

June 7, 2005

OIG conducted this audit to evaluate controls over the use of NRC telecommunications services and the physical security of NRC telecommunications systems. OIG found that the agency needs to strengthen controls over the use of telecommunications services and the physical security of NRC telecommunications systems.

Open Recommendations	Actions Pending
3. Revise Management Directive 2.3 and Handbook, "Telecommunications," to include effective management controls over NRC headquarters staff use of agency telecommunications services.	On March 31, 2009, the Office of Information Services (OIS) advised OIG that it was working towards submitting MD 2.3 and its associated handbook to the Office of Administration (ADM). Current estimates call for concurrence in mid-November 2009, with submission to the directives team in December 2009. The goal is to publish the MD and handbook by December 31, 2009.

Audit of the NRC's Decommissioning Program (OIG-05-A-17)

September 21, 2005

OIG conducted this audit to determine whether the NRC's decommissioning program achieves desired performance results, as stated in the Strategic Plan and reported in the Performance and Accountability Report. The audit found that, while the NRC's decommissioning program has processes in place to monitor, evaluate, and report on performance, some performance results could not be verified. In addition, although staff implemented most of the recommendations from an FY 2003 self-evaluation of the program, the agency had not made progress on a few recommendations.

Open Recommendations	Actions Pending
1. Clarify and disseminate expectations for generating and maintaining supporting documentation for performance data to staff responsible for preparing and collecting performance data.	Revised MD 4.7 and Handbook will clarify expectations for generating and maintaining supporting documentation for performance data.
	In August 2007, the Commission directed the Chief Financial Officer, in coordination with staff, to provide options for improving the agency's budget formulation process. The staff developed and implemented a new top-down budget process in formulating the FY 2010 and FY 2011 budgets. Subsequently, the staff considered lessons learned from the NRC task force that reviewed the agency's budget formulation process.
	As a result of lessons learned from the budget process, the agency decided to address its strategic planning process and performance assessment in separate management directives. MD 4.7 and Handbook will address the roles and responsibilities in the agency's budget formulation process.
	Based on the task force's current schedule for issuing guidance on the agency's budget formulation process, the staff expects to publish MD 4.7 and Handbook in January 2010.

System Evaluation of Security Controls for Stand-alone Personal Computers and Laptops (OIG-05-A-18)

September 22, 2005

OIG conducted this evaluation as part of its review of the NRC's implementation of FISMA for FY 2005, with the objectives of evaluating the effectiveness of NRC security policies, procedures, practices, and controls for stand-alone personal computers (PCs) and laptops. The audit found that these policies, procedures, practices, and controls were not adequate, that the devices were not monitored for compliance with Federal regulations, and that agency information technology (IT) coordinators' understanding of disposal practices for these devices was not consistent.

Open Recommendations	Actions Pending
2. Develop and require users to sign a rules of behavior agreement accepting responsibility for implementing security controls on stand-alone PCs and laptops.	The NRC developed its agencywide rules of behavior for authorized computer use and provided them as part of the annual computer security awareness course. As part of completing the course, users are required to electronically acknowledge the rules of behavior.
3. Develop and implement procedures for verifying all required security controls are implemented on stand-alone PCs and laptops.	NRC staff is currently reviewing the compliance review process. This process requires meeting and auditing at either the system or program-office level (or both) to discuss specific agenda items regarding a system, common issues that impact a number of systems, and verification of security controls for PCs and laptops. The Computer Security Office (CSO) expects to finalize the process in the first quarter of FY 2010 and begin a new quarterly compliance review process in the second quarter of FY 2010.
4. Provide users guidance on compliance with Executive Order 13103, "Computer Software Piracy," for stand-alone PCs and laptops.	The agency developed and disseminated clear guidance on compliance with Executive Order 13103 for stand-alone PCs and laptops as part of the standard rules of behavior discussed above under Recommendation 2.

The standard rules of behavior include statements regarding compliance with Executive Order 13103 for stand-alone PCs and laptops. As part of the agency's computer security awareness course, participants electronically acknowledge the rules of behavior. |
| 5. Develop and require users to sign a rules-of-behavior agreement acknowledging their compliance with Executive Order 13103, "Computer Security Piracy," for stand-alone PCs and laptops. | As part of the development of the standard rules of behavior discussed above under Recommendations 2 and 4, the agency developed a standard rules-of-behavior agreement for users to acknowledge their compliance with Executive Order 13103 for stand-alone PCs and laptops. NRC offices were notified of the requirement for all users of such devices. |

6. Develop and implement procedures for monitoring compliance with Executive Order 13103, "Computer Security Piracy," for stand-alone PCs and laptops.

Procedures for monitoring compliance with Executive Order 13103 for stand-alone PCs and laptops will be developed and issued as part of the standard rules of behavior discussed above under Recommendation 2

The compliance review process is currently undergoing coordination and review. This process requires meeting and auditing at either the system or program office level (or both) to discuss specific agenda items regarding a system, common issues that impact a number of systems, and verification of security controls for PCs and laptops. CSO expects to finalize the process in the first quarter of FY 2010 and begin the new quarterly compliance review process in the second quarter of FY 2010.

Evaluation of the NRC's Use of Probabilistic Risk Assessment In Regulating the Commercial Nuclear Power Industry (OIG-06-A-24)

September 29, 2006

The objectives of this evaluation were to determine whether the NRC is following prevailing good practices in probabilistic risk assessment (PRA) methods and data in its use of PRA, using prevailing good practices in PRA methods and data appropriately in its regulation of nuclear power plant licensees, and achieving the objectives of the PRA policy statement. The evaluation concluded that, although the NRC is employing prevailing good practices in regulation of nuclear power plants, the agency lacks formal, documented processes and associated configuration control for PRA computer models and software.

Open Recommendations	Actions Pending
3. Conduct a full verification and validation (V&V) of the Systems Analysis Program for Hands-On Integrated Reliability Evaluations (SAPHIRE) Version 7.2 and Graphical Evaluation Module (GEM). (SAPHIRE and GEM are software programs used to perform evaluations of Standardized Plant Analysis Risk Model (SPAR) models and provide risk results based on the events or initiators evaluated.)	Because development of SAPHIRE Version 8 is in progress, a full V&V of SAPHIRE Version 7 would not be an effective use of resources. Therefore, the release of SAPHIRE Version 8, which will include an independent verification and validation (IV&V), will close this recommendation.
	The Office of Nuclear Regulatory Research (RES) continues to be on schedule to release SAPHIRE Version 8 in April 2010. Important quality assurance activities have commenced since the last update. For example, IV&V activities began in April 2009. Additionally, the RES staff is performing an internal peer review in accordance with RES Office Instruction (OI) PRM-010, "Peer Review of RES Projects," as well as periodic software audits of SAPHIRE Version 8, in accordance with RES OI PRM-012, "Software Quality Assurance for RES Sponsored Codes." The staff will incorporate the results of these efforts and ongoing beta software testing as appropriate into the development of Version 8 to ensure that the April 2010 release is of sufficient quality to support NRC needs.

Audit of the NRC's Technical Training Center (OIG-07-A-05)

January 9, 2007

This audit identified opportunities to improve the economy, efficiency, and effectiveness of the Technical Training Center's operations.

Open Recommendations	Actions Pending
1. Revise MD 13.1 to require that property inventories should include independent verification of the property by someone other than the property holder.	The revision to MD 13.1, "Property Management," was sent to all offices for comment on April 2, 2009, and ADM resolved the comments received, including those of OIG, in mid-August 2009. MD 13.1 is currently undergoing final ADM review before being forwarded for final agency review and approval. ADM expects to issue revised MD 13.1 by January 29, 2010.
11. Include questions specific to instructor performance on all course evaluations.	By memorandum dated April 20, 2009, OIG stated that it will close this recommendation once the agency develops questions specific to instructor performance and includes those questions on all course evaluations. In September 2009, the Human Resources Training and Development (HRTD) staff revised Operating Procedure (OP)-410, "HRTD Training Evaluation," to include questions specific to instructor performance on all course evaluations.

Audit of the NRC's Regulation of Nuclear Fuel Cycle Facilities (OIG-07-A-06)

January 10, 2007

This audit determined whether the NRC has an effective and efficient approach to fuel cycle facility oversight. The audit found that the NRC could enhance the current Fuel Cycle Facility Oversight Program by developing and implementing a framework modeled after a structured process, such as the Reactor Oversight Process (ROP).

Open Recommendations	Actions Pending
1. Fully develop and implement a framework for the Fuel Cycle Facility Oversight Program (FCFOP) that is consistent with a structured process, such as the Reactor Oversight Process (ROP).	Agency corrective actions include initiatives to improve fuel cycle oversight, including performing a structured evaluation of integrated safety analysis (ISA) annual updates, providing fuel cycle input to a revision of the NRC enforcement policy, and completing a safety culture pilot plan. The staff has completed the review of the 2007 ISA annual updates and has developed changes to the review process. The ISA update review was concluded following the review of the 2008 annual updates. The staff has drafted proposed changes to the NRC enforcement policy to align the policy with revisions to Title 10 of the *Code of Federal Regulations* (10 CFR) Part 70, "Domestic Licensing of Special Nuclear Material." The enforcement policy revision will conclude when the staff issues the new policy at the end of 2009. The lengthiest corrective action is the two-phase Office of Nuclear Material Safety and Safeguards safety culture project plan, of which Phase I is complete. Phase II of the plan consists of implementing the Phase I results. The staff plans to incorporate the Phase I results into the new FCOP. The FCOP project is currently underway with the formation of an FCOP working group and completion of a number of public meetings with industry.

Audit of the NRC's Badge Access System (OIG-07-A-10)

January 23, 2007

This audit determined whether the current badge access system meets its required operational capabilities and provides for the security, availability, and integrity of the system data.

Open Recommendations	Actions Pending
13. In accordance with NRC requirements for listed systems, develop an access system security plan and appoint an information system security officer.	ADM received several security categorization documents for updating to newer templates, causing a delay in the process. Since Access is a listed security system on a fully enclosed network, the OIS contractor did not give this task a high priority, causing additional delay. Once approved, the staff will forward the security categorization documentation, which officially lists the information system security officer (ISSO) for Access, with the remainder of the certification and accreditation (C&A) documentation. An ISSO and alternate ISSO have been appointed (ADAMS Accession No. ML091200613). Approval of the Access system security plan is expected by November 2, 2009.
15. Complete the actions necessary to address the access weaknesses contained in the penetration test reports.	ACCESS is on a fully enclosed network environment and does not connect to any other system or to the Internet. Because of other high priorities, ADM has determined that it is not cost effective or imperative to correct the findings from the penetration tests with the current, closed network, since the implementation of Homeland Security Presidential Directive (HSPD)-12 will result in system upgrade or replacement. Many of the findings were related to weaknesses present only if the system is connected to other systems or to the Internet. ADM is working with CSO to ensure that any issues which may arise during system upgrade are immediately resolved. This is part of the C&A process. Recommendation 15 is currently scheduled to be resolved by December 29, 2009.

Audit of the NRC's Noncapitalized Property (OIG-07-A-14)

July 12, 2007

This audit determined whether the NRC has established and implemented an effective system of management controls for maintaining accountability and control of noncapitalized property.

Open Recommendations	Actions Pending
7. Modify MD 13.1, "Property Management," to reference, where applicable, MD 12.5, "NRC Automated Information Security Program," to include procedures for coordinating with OIS regarding missing property that contains or may contain personally identifiable information (PII).	The revision to MD 13.1 was sent to all offices for comment on April 2, 2009, with comments due in 30 days. Comments were received from the offices, including the Office of the Inspector General, and were resolved by the Office of Administration (ADM) by the middle of August 2009. The requisite changes were made to the record copy of MD 13.1 and final ADM review was completed on October 29, 2009. The package is being forwarded for final agency review and approval. It is anticipated that the revised MD 13.1 will be issued by January 29, 2010.
11. Work with the OIG to modify MD 13.1 to develop a process for notifying the OIG Assistant Inspector General for Investigations of all reports (Form 395, "Report of Property for Survey") of missing sensitive property and missing nonsensitive property with a current value of at least $1,000.	The revision to MD 13.1 was sent to all offices for comment on April 2, 2009, with comments due in 30 days. Comments were received from the offices, including the Office of the Inspector General, and were resolved by the Office of Administration (ADM) by the middle of August 2009. The requisite changes were made to the record copy of MD 13.1 and final ADM review was completed on October 29, 2009. The package is being forwarded for final agency review and approval. It is anticipated that the revised MD 13.1 will be issued by January 29, 2010.

Audit of the NRC's License Renewal Program (OIG-07-A-15)

September 5, 2007

OIG conducted an audit of the license renewal review program and, while acknowledging the existence of a comprehensive license renewal review process, the audit identified several areas in which improvements would enhance program operations. The audit included eight recommendations. Based on prior staff responses to the audit recommendations, OIG closed Recommendations 1, 2, 3, 5, 6, and 8, and determined that Recommendations 4 and 7 will be resolved once the staff finalizes license renewal guidance documents. The staff committed to update the guidance documents by June 30, 2009. On September 24, 2009, the Office of Nuclear Reactor Regulation issued a status memorandum to OIG confirming issuance of the guidance documents.

Open Recommendations	Actions Pending
4. Establish requirements and management controls to standardize the conduct and depth of license renewal operating experience reviews.	This recommendation was previously considered resolved and will close once OIG completes a review of the guidance documents submitted on September 24, 2009, to determine whether the actions provided fully satisfy the intent of the recommendation. The guidance documents submitted for OIG review include the License Renewal Project Manager Handbook Attachment 29, Revision 1, "Safety Evaluation Report Writing Guidelines and Samples"; the License Renewal Project Management Handbook Attachment 31, "Operating Experience Review Responsibilities"; and the License Renewal Project Manager Handbook Attachment 32, "Audit Report Guidance."
	These guidance documents establish requirements and management controls to standardize the conduct and depth of license renewal operating experience reviews.
7. Establish a review process to determine whether or not Interim Staff Guidance (ISG) meets the provisions of 10 CFR 54.37(b), and document accordingly.	This recommendation was previously considered resolved. It will close once the OIG completes a review of the license renewal interim staff guidance (LR-ISG) process document submitted on September 24, 2009, to determine whether the actions provided fully satisfy the intent of the recommendation. The staff issued a revision to the LR-ISG process to include provisions to determine and document whether the requirements of 10 CFR 54.37(b) apply to an LR-ISG.

Review of the NRC's Process for Placing Documents In the Agencywide Documents Access and Management System Public and Nonpublic Libraries (OIG-07-A-16)

September 6, 2007

This audit determined the effectiveness and consistency with which the staff profiles and processes documents for entry into the public or nonpublic Agencywide Documents Access and Management System (ADAMS) libraries.

Open Recommendations	Actions Pending
3. After MD 3.4 and supporting guidance are updated and consolidated, conduct a training needs analysis and develop appropriate training for staff with responsibilities for determining whether ADAMS records should be publicly or nonpublicly available.	The NRC Professional Development Center offers an existing ADAMS training program. OIS is working with Office of Human Resources to develop additional course materials on determining which ADAMS documents should be publicly available and which should not. Once implemented, this will help staff make informed decisions. A training needs analysis was conducted and training for staff with responsibilities for determining whether ADAMS records should be publicly or nonpublicly available has been included in the course, "ADAMS Document Processing." Staff will report on this in a status report due to OIG on November 30, 2009.
6. Conduct periodic assessments of the accuracy with which NRC staff are applying the agency's criteria for designating records as public or non-public by assessing a random sample of records against the agency's criteria for making these determinations.	The agency issued MD 3.4, "Release of Information to the Public," on February 6, 2009. In October 2009, OIS will conduct an annual assessment of the accuracy with which the staff applies the agency criteria for designating records as public or nonpublic by evaluating a random sample of records against the agency criteria for making these determinations. The results of the initial assessment will be reported in the status report due to OIG on November 30, 2009.

Audit of Assessment of Security at NRC Buildings In Rockville, MD; Bethesda, MD; and Las Vegas, NV (OIG-07-A-18)

September 25, 2007

These security assessments determined the adequacy of physical security and emergency planning measures at the identified NRC buildings.

Open Recommendations	Actions Pending
11. Post signs near vehicle entrance directing pedestrians further west along Marinelli Avenue, and paint "Crosswalk" to direct pedestrians along a safe path to two controlled entry points.	Implementation of HSPD-12 included an overall assessment of physical access controls at the NRC headquarters complex. An NRC consultant completed an assessment of Recommendation 11 on February 29, 2008. Based on that assessment, the staff is preparing a proposed plan and cost analysis on installing a security fence to enclose the rear of the complex. The fence will control pedestrian traffic entering the One White Flint North and Two White Flint North buildings at the P1 levels. Because of the complexity of the terrain and associated easements with the NRC property, the agency awarded an architectural and engineering contract to Oudens & Knoop on September 26, 2008. Oudens & Knoop completed the design phase of this project in August 2009. The NRC must brief the Maryland National Capitol Planning Commission on the security fence project. This briefing is tentatively scheduled for November 2009. The construction phase of this project is tentatively scheduled for spring 2010. Recommendation 11 is scheduled for completion by June 30, 2010.

Independent Evaluation of the NRC's Implementation of the Federal Information Security Management Act for FY 2007 (OIG-07-A-19)

September 28, 2007

An independent evaluation of the NRC's implementation of Federal Information Security Management Act (FISMA) for FY 2007 found that the NRC information security program needed improvements.

Open Recommendations	Actions Pending
11. Develop and implement quality assurance procedures for the Plan of Action and Milestones (POA&Ms).	In addition to documenting the procedures, CSO will also undertake other steps related to improving the quality of POA&M information. This will include (1) documenting procedures for conducting independent verification and validation of POA&Ms to ensure their adequacy as part of the security assessment review process (2) acquiring additional contract support to assist in establishing a compliance review process in which CSO will review security documentation, conduct vulnerability scanning, and (3) meet with each system owner on an annual basis to verify the status of remediation efforts; to assess the comprehensiveness of planned corrective action; and to validate the accuracy of tasks, responsibilities, and milestones for each outstanding weakness.
	These activities will take place quarterly and target approximately 25 percent of the overall number of POA&Ms.
	After 6 months of research and evaluation, CSO selected Xacta as the agency's tool for automating the POA&Ms. CSO recently purchased the Xacta application. Additionally, CSO developed the POA&M process (ADAMS Accession No. ML092810195) and plans to start meeting with the system owners in the first quarter of FY 2010.

14. Develop and implement procedures for ensuring that employees and contractors with significant IT security responsibilities are identified, that they receive security awareness and training, and that the individual and associated training are readily correlated. This is Recommendation 10 from OIG-05-A-21, "Independent Evaluation of NRC's Implementation of FISMA for Fiscal Year 2005."

All IT security roles were identified by means of a data call issued July 31, 2009, by CSO to all offices. The IT Security Role-Based Training Plan, available at http://www.internal.nrc.gov/CSO/documents/FINAL%20IT%20Role-Based%20Training%20Plan.doc, identifies the training requirements for those with significant IT responsibilities, the type of training expected for each role, and the frequency of training for each role. CSO is working with a contractor to draft and present six specific role-based courses which address six roles (ISSO, system administrator, IT manager, system owner, executive manager, and senior manager). Two ISSO courses are planned for the second quarter of FY 2010. The system owner is responsible for using the training plan procedures to address the training needs of his or her personnel with IT roles.

Audit of the NRC's Alternative Dispute Resolution Program (OIG-08-A-03)

December 14, 2007

This audit was conducted to determine whether the enforcement-related alternative dispute resolution (ADR) program, both early and postinvestigation ADR, was complete and ready for full implementation. The NRC deemed the ADR pilot program a success, and the staff, ADR participants, and other external stakeholders expressed satisfaction with the program. However, OIG found that the postinvestigation ADR process was not ready for full implementation because of weaknesses in the program's guidance and management controls.

Open Recommendations	Actions Pending
2. Incorporate the interim guidance into the Enforcement Policy and Manual.	The staff is preparing to submit a major revision of the Enforcement Policy, including the policy related to ADR, to the Commission. The agency published the draft Enforcement Policy in the *Federal Register* in September 2008, and republished the revised Enforcement Policy supplements in June 2009. After resolution of both internal and external stakeholder comments, the staff will submit the policy to the Commission for approval by December 31, 2009. The Enforcement Manual revision is complete.

Audit of the NRC's Planned Cybersecurity Program (OIG-08-A-06)

March 18, 2008

This audit determined how upcoming changes to the NRC's cybersecurity oversight processes might impact the agency's physical security inspection program.

Open Recommendations	Actions Pending
1. Develop and implement plans for a cybersecurity oversight program that captures skill set and workload requirements for cybersecurity inspections, and targets resources to prepare for program implementation in calendar year 2010.	Through the planning, budgeting, and performance management process, the staff has requested staff and contract resources for program development in calendar year 2010. With the requested resources, the staff tentatively plans to develop a temporary instruction inspection procedure and related enforcement guidance, conduct training for the cybersecurity inspection team, and prepare for associated industry workshops. The inspections conducted using the temporary instruction will provide the framework for further development of the cybersecurity oversight program and the program's transition into the ROP.

Audit of the NRC's Oversight of Licensees' Nuclear Security Officers (OIG-08-A-07)

March 18, 2008

This audit assessed the NRC's oversight of security officers employed by licensees to protect nuclear power plants. It considered the fitness-for-duty regulations in 10 CFR Part 26, "Fitness for Duty Program," as well as pertinent regulations in 10 CFR Part 73, "Physical Protection of Plants and Materials."

Open Recommendations	Actions Pending
2. Integrate behavioral observation program regulations with access authorization regulations in ongoing Part 73 rulemaking.	The staff has issued Regulatory Guide (RG) 5.66, "Access Authorization Program for Nuclear Power Plants." RG 5.66 endorses Nuclear Energy Institute (NEI) 03-01, "Nuclear Power Plant Access Authorization Program," which provides the implementing guidance to address this recommendation. The staff provided OIG with information to close this recommendation in a memorandum dated September 16, 2009 (ADAMS Accession No. ML092380191).

Audit of the NRC's Continuity of Operations Plan (OIG-08-A-10)

May 21, 2008

This audit determined NRC's compliance with requirements for security surveys of the NRC's continuity of operations plan facilities.

Open Recommendations	Actions Pending
1. Revise current agency guidance governing security surveys of NRC continuity facilities to reflect Federal requirements (as originally stated in Federal Preparedness Circular 65 and superseded by Federal Continuity Directive 1) regarding annual physical security surveys of continuity facilities.	The revised MD 12.1, "NRC Facility Security Program," reflects the Federal Continuity Directive (FCD) 1, "Federal Executive Branch National Continuity Program and Requirements" requirement to provide for annual physical security inspections of continuity facilities. Because of other agency priorities, this document was not forwarded to the appropriate offices for review and comment as indicated in the memorandum dated May 15, 2009 (ADAMS Accession No. ML091630475). ADM plans to publish the revised MD 12.1 by February 5, 2010.

Audit of the NRC's Accounting and Control Over Time and Labor Reporting (OIG-08-A-11)

Date of Audit

OIG conducted an audit of the NRC's time and labor system on June 17, 2009. The objectives of the audit were to determine whether the NRC established and implemented internal controls over time and labor reporting to provide reasonable assurance that hours worked in pay status and hours absent are properly reported and that the time and labor system is easy and efficient to use.

Open Recommendations	Actions Pending
3. The CFO should conduct a detailed system analysis and eliminate redundant paper forms that are not needed.	The modernization project for the time and labor system is scheduled to be completed by March 2010. As part of this modernization, the Office of the Chief Financial Officer (OCFO) is working to incorporate an electronic workflow process, which would allow for electronic signatures. OCFO has met with the Office of Human Resources to discuss the possible elimination of various leave request forms and has also met with the National Treasury Employees Union. Preliminary findings indicated that the summary approval report, all leave request forms, unit transfer forms, and security request forms can be part of the electronic workflow process.
4. The CFO should ensure the use of electronic signature for time reporting and approval.	The modernization project for the time and labor system is scheduled to be completed by March 2010. As part of this modernization, OCFO is working to incorporate an electronic workflow process, which would allow for electronic signatures.

Evaluation of the NRC's Training and Development Program (OIG-08-A-13)

July 16, 2008

This evaluation determined the effectiveness of the NRC's Training and Development Program to meet current and future needs.

Open Recommendations	Actions Pending
2. Develop a plan and timeline for completing any missing course documentation.	By memorandum dated May 9, 2009, OIG stated that this recommendation will be closed when HRTD provides OIG with the documentation that verifies that HRTD has developed a plan and timeline for completing any missing course documentation. In June 2009, HRTD created a matrix of missing course materials and a plan for completing any missing course documentation.
3. Complete HRTD Operating Procedure 404—Training Material Control to include a standard process for version control, tracking changes, and assigning accountability for changes.	By memorandum dated May 9, 2009, OIG stated that this recommendation will be closed when OIG receives a copy of the completed OP-404, "Training Material Control," that includes a standard process for version control and tracking changes. HRTD will update OP-404 to include a standard process for version control and tracking changes by December 31, 2009.
4. Develop a plan to centralize course materials in one location, preferably a central repository on a shared server.	By memorandum dated May 9, 2009, OIG stated that this recommendation will be closed when HRTD provides OIG with documentation that verifies that HRTD has developed a plan to centralize course materials in one location. In June 2009, HRTD developed and began implementing a plan to centralize its course materials in ADAMS.
7. Within contractual limitations, identify and schedule courses at the same time each year so employees can anticipate the availability of courses. Communicate these courses and dates widely to make sure employees, supervisors, training coordinators, and managers are aware of the annual schedule.	By memorandum dated May 9, 2009, OIG stated that this recommendation will be closed once OIG receives evidence that HRTD identifies and schedules courses at the same time each year and communicates this information to affected individuals. HRTD established a working group and developed an annual, repeating schedule; the schedule was communicated to affected individuals in June 2009.

8. Before implementing the enforcement of course prerequisites in the Learning Management System:

    a. Determine the impact on employees' ability to take all required training within the allotted timeframe.

    b. Communicate the change to NRC personnel in advance and allow opportunities for feedback.

By memorandum dated May 9, 2009, OIG stated that this recommendation will be closed once OIG receives documentation that verifies the following:

(1) OP-0403, "Course Administration," contains the process HRTD uses to determine the impact on employees' ability to take all required training within the allotted timeframe.

(2) HRTD has notified all office training coordinators of existing prerequisites and allowed NRC personnel opportunities for feedback.

In reference to Item a, the process for identifying prerequisite course requirements was incorporated into OP-0403 in June 2009.

In reference to Item b, the iLearn learning management system (LMS) was searched and 15 courses were identified that required completion of prerequisite courses. A reminder e-mail listing the 15 courses was forwarded to the office and regional training coordinators in March 2009.

9. Develop and implement new performance metrics to demonstrate mission alignment, effectiveness, and efficiency.

By memorandum dated May 9, 2009, OIG stated that this recommendation will be closed when HRTD develops and implements new performance metrics to demonstrate mission alignment, effectiveness, and efficiency and provides OIG with documentation that verifies that this action has been taken.

The staff has recently developed an agencywide approach for measuring the effectiveness of training. The staff is piloting and evaluating this method, after which, the staff will have adequate data to evaluate, develop, and implement any necessary new performance metrics. The staff plans to begin using the new method for measuring training effectiveness in the first quarter of FY 2010 and will evaluate appropriate performance metrics in the second quarter of FY 2010.

10. Develop and implement a plan to leverage the capabilities of the LMS for collection and reporting of chosen metrics. Specifically, evaluate the competency model capabilities to determine if they meet NRC's needs, including identifying competencies, linking courses (or course modules and learning objectives) to identified competencies, and closing critical skill gaps.

By memorandum dated May 9, 2009, OIG stated that this recommendation will be closed when HRTD develops and implements a plan to leverage the capabilities of the LMS for collection and reporting of chosen metrics and provides OIG with the documentation that verifies that this action is complete.

HRTD developed and implemented a plan to leverage the capabilities of the LMS for collection and reporting of chosen metrics in September 2009.

11. Develop and implement a comprehensive cost tracking capability (including cost data for each course) to determine the most economical and efficient method to meet NRC's training needs.

By memorandum dated May 9, 2009, OIG stated that this recommendation will be closed when HRTD provides documentation that verifies that OP-406, "Training Program Development Process," has been revised to include specific guidance for applying NRC cost-benefit principles for determining the most economical and efficient method to meet NRC's training needs.

HRTD revised OP-406 in September 2009, to include specific guidance for applying NRC cost-benefit principles for determining the most economical and efficient method to meet NRC's training needs.

13. Develop an evaluation strategy plan that defines the data HRTD needs to collect at varying levels to demonstrate the impact of its programs on the agency.

By memorandum dated May 9, 2009, OIG stated that this recommendation will be closed when HRTD develops an evaluation strategy plan (and provides the appropriate documentation to OIG) that defines the data HRTD needs to collect at varying levels to demonstrate the impact of its programs on the agency.

HRTD will engage its stakeholders to develop an evaluation strategy plan by March 31, 2010.

14. Evaluate the capability for collecting evaluation data via the Learning Management System.

   a. If the Learning Management System's capabilities meet the agency's needs, develop a business case for purchase and deployment of additional capabilities.

   b. If the Learning Management System's capabilities do not meet the agency's needs, develop a plan for using alternative technologies to collect and analyze evaluation data.

16. Develop an implementation plan for e-learning that includes, at a minimum:

   a. An assessment of NRC's baseline technology.

   b. A plan for roll-out, implementation, maintenance and ongoing evaluation of additional Learning Management Systems capabilities. The implementation plan should include a cost/benefit analysis of the available LMS features and how they can support NRC's business needs.

By memorandum dated May 9, 2009, OIG stated that this recommendation will be closed once HRTD provides the results of its evaluation and documentation of either the business case for purchase and deployment of additional capabilities or a plan for using alternative technologies to collect and analyze evaluation data.

The staff will complete evaluation of the capability for collecting evaluation data via the LMS by March 31, 2010.

An implementation plan for e-learning has been drafted and is under review within the NRC. This plan was developed using the results of an assessment of NRC training technologies and their relation to core competencies. This plan envisions that the e-learning courses will be hosted on the LMS.

The NRC is currently upgrading the LMS to Service Pack 5, as well as assessing LMS competency module capabilities.

HRTD has initiated a process to identify the necessary e-learning skills for NRC staff.

17. Develop a process similar to the Strategic Workforce Planning Process to:

    a. Determine the e-learning skill sets necessary to meet NRC's business needs.

    b. Assess the current staff to determine the availability of the needed skills.

    c. Develop a plan to close any identified gaps—either through training, hiring, or outsourcing.

HRTD has developed a catalogue of e-learning tools appropriate for the NRC environment. The next step is to develop and design a Web-based course that incorporates some of the e-learning training methods. A vendor has been selected and course development will be initiated in FY 2010. The NRC staff will participate in course development and gain valuable experience in the conversion from instructor-led classroom training to e-learning. Part of the contract will provide for maintenance of the online courses. Based on lessons learned as these courses are developed, management will decide on the appropriate mix of e-learning solutions. Consideration may include a mix of training for current staff, hiring the necessary skills, or outsourcing the requisite skills.

Audit of the NRC's Premium Class Travel (OIG-08-A-16)

OIG conducted an audit of the implementation of the NRC's premium class travel on September 15, 2008. The objectives of the audit were to determine whether travel costs associated with premium air travel (i.e., per diem) are properly authorized, justified, and documented and to determine whether premium air travel is properly authorized, justified, and documented. OIG specifically assessed compliance with requirements in OMB Memorandum M-08-07.

Open Recommendations	Actions Pending
1. Update Management Directive 14.1 to clearly identify premium travel authorizing officials; clarify "Delegation of Authority" and require this to be in written form; and clarify the 14-hour rule, specifically the rest period.	MD 14.1, "Official Temporary Duty Travel," has been revised to incorporate these changes. The staff is finalizing the various revisions and edits to the MD before it is submitted for formal review and concurrence. OCFO considers Recommendation 1 resolved.
6. As the Nuclear Regulatory Commission (NRC) transitions to eTravel, develop controls that require premium travel to be authorized by the appropriate officials; ensure premium travel justification fully meets Federal Travel Regulations (FTR) criteria; and require travel authorization forms to state reasons why premium travel is required.	When OCFO deploys the premium class feature in eTravel at the end of the first quarter of FY 2010, the travel documents will be routed only to those NRC employees with the authority to authorize premium travel. OCFO considers Recommendation 6 resolved.

Audit of the NRC's Enforcement Program (OIG-08-A-17)

September 26, 2008

The objective of the audit was to review the NRC's enforcement program to determine whether the program is comprehensive and consistently implemented and whether enforcement decisions are based on complete and reliable data. OIG identified that the regional offices implement the enforcement program inconsistently because the agency has not issued clear and comprehensive guidance to facilitate the program. In addition, the audit identified that information used for decisionmaking and reporting purposes is not complete and reliable.

Open Recommendations	Actions Pending
1. Develop guidance establishing expectations for dispositioning violations and the participants needed for enforcement decision-making.	The NRC staff is currently revising the Enforcement Manual and Inspection Manual Chapter (IMC) 2800, "Materials Inspection Program," to specify the minimum level of review and concurrence necessary for approving nonescalated enforcement actions.
2. Define data collection requirements for non-escalated actions. 3. Develop a quality assurance process to ensure that enforcement data is accurate and complete.	The NRC staff is currently developing a Web-based licensing system that will track nonescalated enforcement actions issued to materials licensees. The staff has evaluated the capabilities available with the reactor program system and determined that it is a sufficient tool for tracking and trending nonescalated reactor enforcement actions.

Implementation of the Federal Information Security Management Act for FY 2008 (OIG-08-A-18)

September 26, 2008

The objective of this review was to perform an independent evaluation of the NRC's implementation of FISMA for FY 2008.

Open Recommendations	Actions Pending
1. Update the U.S. Nuclear Regulatory Commission (NRC) System Information Control Database to identify all interfaces between systems.	On June 15, 2009, CSO, in coordination with OIS, ensured that the NRC System Information Control Database had been updated to identify interfaces for systems listed in the inventory.
2. Develop and implement procedures to ensure interface information in the NRC System Information Control Database is consistent with interface information in security plans and risk assessments.	CSO completed development and, on June 15, 2009, implemented procedures for ensuring that security information and interfaces are consistent with information in corresponding system security plans and risk assessments.
3. Develop agency-wide policy and procedures regarding the implementation and monitoring of Federal Desktop Core Configuration controls for all desktop and laptop computers, including both those that are centrally managed under the agency's seat management contract and those that are owned by the agency regardless of whether or not they are connected to the agency's network.	An agencywide policy and associated procedures for implementation and monitoring of Federal desktop core configuration (FDCC) controls for all desktop and laptop computers have been developed. CSO, in coordination with OIS, has developed the following: – configuration standards for NRC. – guidance for general laptops. – procedures for applying critical updates to safeguards information (SGI) laptops. – an SGI standalone listed system minimum security checklist to ensure appropriate laptop configuration.

– standard system security plans for NRC laptops.

– laptop security policy provided.

Additionally, all computers connected to the NRC network receive FDCC settings through the use of group policy object objects settings. Computers that are not attached to the network are loaded with these controls as part of the standard configuration image, and additional controls are implemented through local security policy. These controls were implemented June 15, 2009.

4. Develop a process for verifying that all Federal Desktop Core Configuration controls are implemented for all desktop and laptop computers, including both those that are centrally managed under the agency's seat management contract and those that are owned by the agency regardless of whether or not they are connected to the agency's network.

The NRC has deployed the security content automation protocol scanners to verify that the agency is compliant with M-08-22, "Guidance on the Federal Desktop Core Configuration (FDCC)," during the system certification and accreditation process. CSO is currently fielding its information assurance system (IAS) to provide real-time assessment of FDCC compliance for networked computers as part of its continuing monitoring assurance activities. Standalone systems are configured to FDCC standards during computer buildout. This recommendation is partially closed pending the completion of the IAS, which is necessary for the NRC to provide agencywide, real-time FDCC assessments. The system is currently scheduled to be completed by September 30, 2010.

Audit of the NRC's Laptop Management (OIG-08-A-19)

September 30, 2009

The audit evaluated the management of laptops, including the effectiveness of the NRC's security policies for laptop computers.

Open Recommendations	Actions Pending
3. Provide mandatory formal training to all IT coordinators and property custodians on how to update security controls on laptops.	Mandatory formal training was provided to all system ISSOs on how to update security controls on laptops. Courses were conducted on June 2, 2009; July 1, 2009; July 8, 2009; and August 7, 2009. A copy of the course slides was provided to those ISSOs unable to attend the training in person. The course slides are available on the CSO Web site. Eighty-five percent of ISSOs completed the mandatory training.
4. Develop a process for verifying that all required security controls are implemented on agency-owned laptops.	The compliance review process is currently going through coordination and review. This process requires meeting and auditing at either the system or program-office level (or both) to discuss specific agenda items regarding a system, common issues that impact a number of systems, and verification of security controls for PCs and laptops. CSO expects to finalize the process in the first quarter of FY 2010 and begin the new quarterly compliance review process in the second quarter of FY 2010.
5. Develop a protocol to facilitate the efficient and routine updating of agency-owned laptops located at headquarters.	OIS has taken steps towards centralized management of all agency laptop computers. OIS now provides OIS-managed laptops to headquarters staff, which are configured for use either outside the agency (e.g., travel or work at home) or as a device that can attach to the production network. Laptops that are authorized to directly attach to the NRC network are updated via the NRC-managed infrastructure, and the travel laptops are configured for daily updates. OIS also assists headquarters staff in ensuring that their office-provided laptops meet current security guidelines.

NRC Director of the Division of Reactor Projects Len West shows end of cycle results during a public meeting discussing Farley Nuclear Plant, near Dothan, AL.

Summary of Financial Statement Audit and Management Assurances

Photo Courtesy of NRC Photo Library

Millstone Power Station is located near New London, CT. It is operated by Dominion Nuclear Connecticut, Inc.

Photo Courtesy of NRC Photo Library

Commissioner Svinicki visits Fermi Unit 1 Containment in Michigan with Mark Satorius, Jim Beall, and licensee personnel in November 2008.

Summary of Financial Statement Audit and Management Assurances

SUMMARY OF FINANCIAL STATEMENT AUDIT

Audit Opinion—Unqualified

Restatement—No

Material Weaknesses—No

SUMMARY OF MANAGEMENT ASSURANCES

Effectiveness of Internal Control over Financial Reporting (FMFIA § 2)

Statement of Assurance—Unqualified

Material Weaknesses—No

Effectiveness of Internal Control over Operations (FMFIA § 2)

Statement of Assurance—Unqualified

Material Weaknesses—No

Conformance with Financial Management System Requirements (FMFIA § 4)

Statement of Assurance—Systems Conform to Financial Management System Requirements

Nonconformance—No

Compliance with Federal Financial Management Improvement Act (FFMIA)

	Agency	Auditor
Overall Substantial Compliance	Yes	Yes
1. Systems Requirements	Yes	Yes
2. Accounting Standards	Yes	Yes
3. United States Standard General Ledger at Transaction Level	Yes	Yes

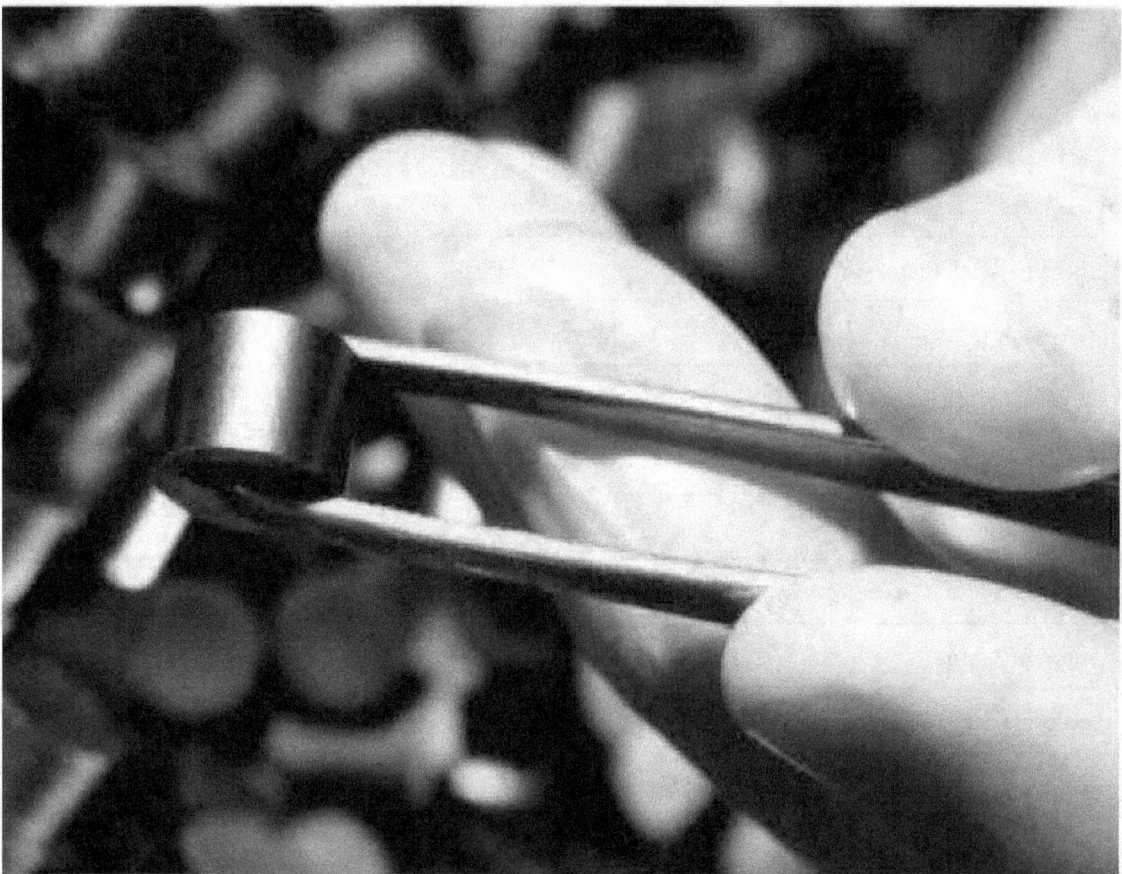

Nuclear Reactor fuel pellet.

Photo Courtesy of NRC Photo Library

Acronyms and Abbreviations

Photo Courtesy of NRC Photo Library

DC Cook Nuclear Plant is located near Benton Harbor, MI. It is operated by Indiana Michigan Power Co.

Acronym	
10 CFR	Title 10 of the *Code of Federal Regulations*
ADAMS	Agencywide Documents Access and Management System
C&A	certification and accreditation
CCDP	Conditional core damage probability
CFO	Chief Financial Officer
CFR	*Code of Federal Regulations*
COL	combined license
CSO	Computer Security Office
CSRS	Civil Service Retirement System
CUI	controlled unclassified information
DOE	U.S. Department of Energy
DOI-NBC	Department of the Interior National Business Center
DOL	U.S. Department of Labor
ECIC	Executive Committee on Internal Control
EDO	Executive Director for Operations
e-Gov	Federal Government's Electronic Government
EPR	Evolutionary Power Reactor
ESBWR	Economic Simplified Boiling-Water Reactor
FCFOP	Fuel Cycle Facility Oversight Program
FECA	Federal Employees Compensation Act
FERS	Federal Employees Retirement System
FISMA	Federal Information Security Management Act
FMFIA	Federal Managers' Financial Integrity Act
FOIA	Freedom of Information Act
FR	*Federal Register*

Acronym	
FY	fiscal year
GAAP	generally accepted accounting principles
GEM	graphical evaluation module
GPRA	Government Performance and Results Act
GSA	General Services Administration
HSPD	Homeland Security Presidential Directive
IAEA	International Atomic Energy Agency
IG	Inspector General
Improvement Act	Federal Financial Management Improvement Act
Integrity Act	Federal Managers Financial Integrity Act
IPSS	Integrated Personnel Security System
ISA	integrated safety analysis
ISG	interim staff guidance
IT	information technology
ITAAC	inspections, tests, analyses, and acceptance criteria
LES	Louisiana Energy Services
MC&A	material control and accounting
MD	management directive
NMMSS	Nuclear Materials Management and Safeguards System
NRC	U.S. Nuclear Regulatory Commission
NSTS	National Source Tracking System
NWF	Nuclear Waste Fund
OBRA-90	The Omnibus Budget Reconciliation Act of 1990
OIG	Office of the Inspector General
OIS	Office of Information Services

Acronym	
OMB	Office of Management and Budget
OUO	official use only
PII	personally identifiable information
POA&M	plan of action and milestones
PRA	probabilistic risk assessment
ROP	Reactor Oversight Process
SAPHIRE	Systems Analysis Program for Hands-On Integrated Reliability Evaluations
SGI	safeguards information

Acronym	
SNM	special nuclear material
SUNSI	sensitive unclassified, nonsafeguards information
T&L	time and labor
TAD	transportation, aging, and disposal
TSP	Thrift Savings Plan
USAID	U.S. Agency for International Development
USEC	United States Enrichment Corporation
V&V	verification and validation

NRC FORM 335 (9-2004) NRCMD 3.7	U.S. NUCLEAR REGULATORY COMMISSION	1. REPORT NUMBER (Assigned by NRC, Add Vol., Supp., Rev., and Addendum Numbers, if any.)
	BIBLIOGRAPHIC DATA SHEET (See instructions on the reverse)	NUREG-1542, Vol. 15

2. TITLE AND SUBTITLE		3. DATE REPORT PUBLISHED	
U.S. Nuclear Regulatory Commission Fiscal Year 2009 Performance and Accountability Report		MONTH	YEAR
		November	2009
		4. FIN OR GRANT NUMBER	
		n/a	

5. AUTHOR(S)	6. TYPE OF REPORT
Richard Rough, et. al	Annual
	7. PERIOD COVERED (Inclusive Dates)
	FY 2009

8. PERFORMING ORGANIZATION - NAME AND ADDRESS *(If NRC, provide Division, Office or Region, U.S. Nuclear Regulatory Commission, and mailing address; if contractor, provide name and mailing address.)*

Resource Management and Support Staff
Office of the Chief Financial Officer
U.S. Nuclear Regulatory Commission
Washington, DC 20555-0001

9. SPONSORING ORGANIZATION - NAME AND ADDRESS *(If NRC, type "Same as above"; if contractor, provide NRC Division, Office or Region, U.S. Nuclear Regulatory Commission, and mailing address.)*

Same as 8, above

10. SUPPLEMENTARY NOTES

11. ABSTRACT *(200 words or less)*

The FY 2009 Performance and Accountability Report provide performance results and audited financial statements that enable Congress, the President, and the public to assess the performance of the agency in achieving its mission and stewardship of its resources. The report contains a concise overview, Management's Discussion and Analysis, as well as performance and financial sections. Details of performance results and program evaluations can be found in Other Accompanying Information.

12. KEY WORDS/DESCRIPTORS *(List words or phrases that will assist researchers in locating the report.)*

Performance and Accountability Report
FY 2009
PAR

13. AVAILABILITY STATEMENT
unlimited
14. SECURITY CLASSIFICATION
(This Page) unclassified
(This Report) unclassified
15. NUMBER OF PAGES
16. PRICE

NRC FORM 335 (9-2004)   PRINTED ON RECYCLED PAPER

AVAILABILITY OF REFERENCE MATERIALS
IN NRC PUBLICATIONS